THE OLD LOG SCHOOL

Gavin Hamilton Green

Natural Heritage/Natural History Inc.

In Memoriam

(On October 13, 1961, Gavin Hamilton Green, of Goderich,
died in his 100th year.)

Now I recall the passing of a friend,
Today, the anniversary of his death,
The hour when from his old lips went the breath
Of mortal life, his journey at an end.

Now I recall the summer that we met
For the first time, when he held out his hand
In kindly greeting; now I understand
Just what it means to care and not forget.

Tis tongue is silent, but his words live on,
Live on in the two volumes that he wrote,
And in the lines of his last cheerful note
To me, that I now fondly gaze upon.

"The Old Log School," "The Old Log House" enshrine
His memories of Huron's early years,
Wherein are mingled with the laughter, tears -
And I am not today ashamed of mine.

Ernest H. A. Home, Strathroy, Ontario
Originally published in the Goderich Signal Star, October, 1962

The Old Log School
Published by Natural Heritage/Natural History Inc.
P.O. Box 69, Station H
Toronto, Ontario M4C 5H7
in conjunction with
The Huron County Historical Society

Design: Derek Chung Tiam Fook
Printed and bound in Canada by Hignell Printing Limited, Winnipeg, Manitoba.

Canadian Cataloguing in Publication Data
Green, Gavin Hamilton, 1862-1961
   The old log school

Rev. & expanded ed. of: The old log school and Huron old boys in pioneer days.
ISBN 0-920474-71-3

1. Green, Gavin Hamilton, 1862-1961.   2. Education – Ontario – Huron County
– History.   3. Frontier and pioneer life – Ontario – Huron County.   I. Title.

LA418.06G7   370.9713'22   C92-093475-7

Contents

IN THE DAYS OF THE OLD LOG SCHOOL

CHAPTER I *30*

A Short History of Our Births, the Old Log School's and Mine

CHAPTER II *36*

Dungannon School – Scenes Amidst Which I Spent Many Happy
Childhood Days

CHAPTER III *58*

Port Albert School, Where I Wore My First Long Pants and Was
Promoted into the Rawhide Whipping Class

CHAPTER IV *61*

Goderich, Tiverton and Sheppardton Schools – At Tiverton About
All I Learned Was to Clean Pigs' Guts for Sausage Casings and
How to Say Naughty Words in Gaelic – At Sheppardton I Enjoyed
Life and Finished My School Education in the Third Book

SKETCHES OF PIONEER DAYS

Acknowledgements

The re-publication of Gavin Green's work, "The Old Log School" is the result of the generosity of the Huron County Historical Society, the Corporation of the Town of Goderich combined with the willing partnership of Barry Penhale, Publisher of Natural Heritage/Natural History Incorporated, and a number of smaller business and individual donations to support the undertaking. The Ontario Heritage Foundation has provided a grant to support research about the life and works of Gavin Green. Several additional items included in the current volume are products resulting from that research grant.

Jayne Cardno, assisted by Robin Wark, gathered archival data and prepared documentation for the biographical profile of Gavin Green and the annotated site list. Jayne Cardno prepared the preliminary drafts of the Introduction. Pat Hamilton and other staff at the Huron County Museum have provided support and willing access to the Huron County Archives whenever corroborating information was required.

The loan of photographs for use in this volume from various sources is acknowledged. The Huron County Museum assisted through the loan of originals and the production of photographic copies as did the staff at the Learning Resources Centre for the Huron County Board of Education. Additional prints and negatives have been provided by Eric McNee, Mrs. Carl Anderson, Carolyn Thompson, Hazel McMichael, 'Bud' McCreath and have been made available through the permission of Marian E. Zinn from "Bush Trails to Present Tales". Much of the original portraiture and a number of the school group pictures are the work of the renowned R.R. Sallows, late of the Town of Goderich, whose photography of pioneer times is well known throughout Canada. The back cover photo is the work of the late Earl MacLaren, also of Goderich. One item in the Green endcover collage is by J.G. Henderson from the Huron County Museum

Collection. The cover photo is from the collection of Eric McNee.

Special note is made to recognize K.K. Dawson, Dungannon. His vivid recollections and historical research have provided important details about the early childhood haunts of Gavin Green and his family in the Dungannon area. The untiring support and indulgence of Janet McCarthy, at the Education Centre from the Huron County Board of Education, Clinton, for assistance in manuscript preparation and numerous clerical support tasks during the process of making arrangements for this re-publication, are also acknowledged.

Paul Carroll, Seaforth, 1992.

Introduction to the Reprint of
Gavin H. Green's "The Old Log School"

In reading "The Old Log School" by Gavin H. Green, it is not long before one wants to sink into a comfortable armchair in front of a crackling wood fire, let the dust settle, find a quilt, and enjoy a good story. This is the ideal way to read Gavin Green's book.

Once caught up in the pages of the "The Old Log School", the reader moves in fifty year time spans between centuries. At one moment, you are reading of events in the 1930s, but before you know it, the 1860s are before your eyes. It is a unique opportunity to explore the change and development of a region and its people over a span of some hundred and thirty years.

Through the pages we meet new, yet old friends; or so they seem to become. The stories flow as does strong tea from a pot shared with friends. They are at once spicy yet sweet-tempered with modesty. Indeed they include all elements of the Gavin Green recipe for writing a book but keep a spontaneity that captures and holds the willing reader. His was a gentle way. He was a man of his time and place and in his desire to record such, was extraordinarily successful.

How many of us remember listening to an elder recount an old tale. We suddenly realize that the special story, once told, may never be heard again. Where was the tape recorder?

Gavin Green had no tape recorder but he recorded his memories on the backs of old calendars. They made great scratch pads. He had a truly remarkable memory. His stories came straight from the heart and all were recounted from his personal experience. They were the accounts of friends and acquaintances of times and places he remembered and does not want forgotten.

Gavin Hamilton Green was "born in the bush" on the 12th Concession of Colborne Township in 1862. His first home was a log house. His father, Peter Hay Green, was a wool carder and later a sawyer in various sawmills. His mother, Janet Kerr, like his father, typified the pioneer. Janet Green managed to cope with the labour of bush life providing everyday comforts of life. She regularly walked to and from Goderich, a day's journey, arriving home with supplies and child in hand to prepare the evening meal.

Gavin Green was a child of the first generation born in Huron. As such, he had the opportunity to grow and develop with the area. The 1860s saw rapid settlement and strong industrial growth in Huron. As young Gavin Green matured, so too did the industry develop to employ him.

In his book, we learn of the various mills, factories and farmers who contracted his labour in places such as Sheppardton, Dungannon and Morristown. Many of these former bustling little communities have disappeared and are unheard of today. None was considered to be an industrial centre. Good jobs, bad jobs, and better jobs; all are described in simple, practical terms; not as the distant past but as though it were only yesterday.

A knowledge of the pioneer days in Huron came to Gavin from his grandfather. Andrew Green had come with the first settlers in 1833 to the Huron Tract. He worked with such figures as Dr. W. "Tiger" Dunlop, a principal founder of the tract and also known as "Lord Warden of the Forests." Andrew Green was the first settler to turn the sod of Colborne Township with a plow. It was through his grandfather that Gavin Green knew the stories of the strong men whom he later profiled in his book.

When he began to write, Gavin Green was an old man by the standard of his time. He was in his seventies or "threescore and ten" he would say. This was an age to which many might aspire but few ever expected to see. Seventy years had allowed Gavin Green to witness many events.

While he did not begin to write until his senior years, ("Long after many my age would be gone", he would likely remark) one senses he had some such goal or purpose early in his life. Certainly he took the advice of a friend who told him to write a book because he had no children with whom to leave his learned wealth.

One senses he felt privileged to have lived so long and to be able to put into print his personal accounts of the development of Colborne Township and the Town of Goderich. He does not hide the fact that he didn't consider himself a great scholar. As you read "The Old Log School", you notice his self-effacing accounts. He graduated from the third reader at the age of eighteen and was a terrible speller. Surviving portions of his calendar scratch pad notes confirm a highly phonetic approach to the printed word and a syntax more suitable for story-telling than for use in a manuscript form.

A strong vein of nostalgia runs repeatedly through Gavin Green's book. This is his testimony, his memorial. Friends are referred to and honoured with respect. Through his anecdotes he says his "good byes" but also he declares an expectation to "see you again in the new world." He gives no impression of gloom. His words are full of expectation for a different future where he will meet again with old friends – the old boys of Huron.

A sense of the era comes to life through his colourful turn of phrase. In describing women's dress, for example, he writes that they wore bustles on their west end when facing East. Such modesty, wit and subtle humour!

He had a folksy style. It is simple and unencumbered, characterized by unique turns of phrase that captured his homespun wit. His straight-forward way of story-telling provides a splendid example of folk writing.

Suitably Gavin Green begins his book with the story of the formation of the first school in Colborne on his family's property. He was well qualified to write of school days, having attended seven schools with a total of fourteen teachers. A student's life is recounted in all its glory and it would be difficult to find more interesting accounts of early school days. Memories of a similar education are likely to be brought to mind by the rich anecdotes and references made to the old school readers. Don't expect a puritanical account. Mr. Green gives a straight-forward, well-rounded account of that life. There is wit, candor and even timeless nonsense. He includes anecdotes from the serious side of things – and in the next line tenders a hasty rebuff to those who are too grim. A lifetime of learning, of personal memories form the nucleus for his original edition of "The Old Log School," privately published in 1939.

Gavin Green was not afraid to express his opinion and his stories provided him with an opportunity to be an advocate for the underdog. In his lighthearted account of the Governor General's 1933 visit to Goderich, we hear of his friend, the local shipbuilder Bill Forrest, whose legal battle with the government is cunningly put to the fore. He was politically astute. One of his best known stories is about the Old Town Clock. In his sardonic sense of humour, he takes on the personality of the clock and bemoans the factors that allowed the clock to lose face. Fact and personal opinion are interwoven in a humorous and lighthearted fashion. He was a bit of a tease.

There was a certain eccentricity to Gavin Green's life. Early in his book he describes his survival from encounters with life's three major dangers – fire, water and alcohol. He felt charmed to have survived childhood and in the spirit of existentialism credits the phenomenon of having been born with a caul cap (the birth membrane still attached, an old belief signalling luck) that he indeed was to lead an enchanted existence.

In 1880, at the age of eighteen, having completed his formal education, Gavin Hamilton Green left Colborne Township to go West. Many young men felt the pull to this newly available and unexplored land on Canada's new frontier.

Through his writings we learn that Gavin Green was stringing wire between Philadelphia and New York for the Bell Telephone Company in 1880. Five years later he was serving on the passenger and cargo vessel "Alberta", a story told in his second book "The Old Log House" published in 1948 and long out-of-print.

Manitoba would wait a little longer, as next, he went North to work on railway construction between Gravenhurst and North Bay. Gavin also worked on lock construction at Magnetawan before returning home to the sawmill town for Christmas.

In 1886, Gavin Green writes that he was back in the wire stringing business with the Postal Telegraph Company in Saginaw, Michigan.

Back in Morristown or Sheppardton in 1888, his younger brother David left for Australia. At that time, Gavin had once more been working at the local Morris mill. He tells us he put out the fire in the boilers for the last time on March 31, 1888 and left for Manitoba.

There is little information about his life in Manitoba but it is believed he farmed near Carberry. In 1892, he returned to Huron County to marry Aggie (Agnes) Bogie, the daughter of Captain Andrew Bogie and the former Martha Sallows. The Bogie family arrived in Huron about the same time as the Green family and had many ties to the early history of the area. Today, a cemetery stone in the Colborne Cemetery near Saltford commemorates the Green's sixty-four years of marriage which began in the gay nineties.

The couple chose to go West and it was there that Gavin Hamilton Green developed a love for things of the past. Perhaps being away made him more sentimental or appreciative of the times at home, as interest in Huron County appeared to increase considerably during this period in his life.

Those were interesting times in the West. The challenges raised by Louis Riel and the skirmishes of the North-West Rebellion were current events, involving great intrigue.

Gavin opened an antique store in Carberry, Manitoba. Tragically, a cousin tells us, it was destroyed by fire shortly after this business venture was started.

Returning home to Goderich in 1902 he established "Ye Olde Curiosity Shoppe" on Hamilton Street, a site quickly discovered by collectors and history enthusiasts. Did he choose a street by the same name as his own on purpose? One can only wonder. Among the interesting objects he displayed in his shop was a diary from the North-West Rebellion. Later he would donate it to Herb Neill, historian friend and founding curator of the Huron County Museum.

For nearly half a century, Gavin Hamilton Green operated "Ye Olde Curiosity Shoppe." (On his hand-printed signs, the name read "Yea Old Curiosity Shop.") People seeking local information were frequently directed to his shop in the expectation that Green would provide reasonable and reasoned answers to their many questions. Many an unmarked grave was pointed out by this man who proffered the facts of history in grand oral tradition. He raised a controversy in the local paper with a story he published in it claiming that Victoria Park, today Judith Gooderham Park, was in fact one of the earliest pioneer cemeteries. His own wife's mother and father were at rest there, he claimed.

Behind the dingy, clouded windows of "Ye Olde Curiosity Shoppe" were concealed a multitude of items. All exhibited their special features for the duration of their sojourn in the shop through layers of accumulated dust. No apologies were made for the dust. A bold sign posted on entry justified its presence. His chapter "Notice to Dust Kickers" adds a humorous glance at his philosophy and the shop. Dust was as much a link to the past and to the rest of the world as anything else in his collection. He reasoned that dust was the tangible evidence of past kings, pharaohs and princesses, perhaps that of our own ancestors. Why rub away these motes of all that has gone before?

Ye Olde Curiosity Shoppe stood on the right side of Hamilton Street as you travel to the Goderich Town Square. Originally it was a two-storey clapboard building. Photos taken between 1902 and 1925 show a grandfather clock and spinning wheel on the outside, mounted to the second storey. This building burnt in September of 1925.

By 1927, Gavin Green had moved into a single storey brick building on the same side of the Street (currently 77 Hamilton Street). Photos of this shop show huge numbers, the digits 66, repeated across the front of the store. Their presence has created much speculation as to their significance. Certainly no passerby could miss them. While some people have suggested that Green had a particular interest in numerology, even spiritualism, others simply suggest that the numerals were posted for good luck. It has also been pointed out that he was approaching age 66 when he began business in that particular building and perhaps he simply celebrated this attainment. One citizen believes he remembers the number '88' being added when Gavin reached that age. Is all of this wishful speculation however? Is it possible that the simple coincidence of his fascination of double digit numbers gives rise to these whispers of mysticism? We note also that he passed "to the Home Land" at age 99.

"He was very superstitious and so many things meant an omen to him", writes a cousin. Coming home one night she tells of meeting him at a street corner. She was carrying a birthday cake. When Gavin found out that the birthdates of his cousin and his wife as well as her sister each fell within six calendar days of the others he became quite excited. "Here was an omen," and certainly there had to be some hidden meaning associated with this coincidence.

Gavin Green was a very complex individual. A darker side of him which very few knew was an obsessive interest in spirits and spirituality. It is with reluctance that friends disclose this peculiar eccentricity. It was an interest that only close friends will acknowledge, but even then with great reservation. Yet it is a tribute to the complexity of the man they accepted as their true and personal friend.

A few of his thoughts describing spiritual experiences are

scribbled on scrap paper in a nearly indecipherable fashion. There is a sense of Houdini (1874-1926) in his notes. His interest extends beyond the then popular notions of spirits in the after-life. Gavin Green's notes show he was familiar with and curious about native American aboriginal spiritualism.

In particular, he mentions the spirit "Niagara", who as a native god could wield the powers of the river now known by the same name. Green saw such spirits as positive forces, life-long accomplices to assist and support in one's life struggles.

Gavin would undertake trance-like meditations while his wife Aggie played the piano. While some may find this aspect of his behaviour rather alarming, it seemed to fit his temperament and his style. He saw spirituality as a means of gaining a greater understanding of life and as a way to help others do the same.

In appearance, Green was an imposing figure, towering more than six feet in height. Many who knew Gavin Hamilton Green personally have commented on his striking appearance. A gentleman of the old school, Green is remembered as a man with whom children did not want to tangle. "You didn't play the more usual tricks on Mr. Green. He commanded a certain respect. He gave pennies to me as a (welcomed) bribe," said one business man, "in order to stop me playing with the organ and the bells in his shop."

He was quite a 'character' say many. He was unusual, not just because he ran a second hand store but because of his passionate attachments to the countless artifacts therein. The shop was in truth a museum of local heritage and its owner an historian of encyclopedic proportions.

Harry J. Boyle, in his London Free Press column, recounted how he "acquired a love for poking through the artifacts of yesterday in Gavin H. Green's Old Curiosity Shop in Goderich." He went on to say how many considered him "a true collector who would spend hours telling you the history of each item."

Gavin Green was a keen witness to his place and time and

we are indebted to his foresight in recording a multitude of happenings from throughout his long life in Huron County. While he had a natural interest in history, his motivation for writing remains open to conjecture. Perhaps it was purely his love of the area and his gift to the people. Perhaps it was a subtle desire for recognition. In the archive collection at the Huron County Museum rest the remnants of his writing. Several hundred pages of unpublished anecdotes pencilled largely on the backs of calendar pads, recount his observations, his recollections and express opinion in his colourful, folksy style.

The 1930s fostered a sense of nostalgia for him. During the lingering Depression many children of the original pioneer families had left home in search of futures elsewhere. Such migration led to disruption and a loss of community but Homecoming Days or Old Boys Reunions in such centres as Goderich, allowed many to keep in touch with their Huron roots. The exchange of early childhood reminiscences was encouraged in this way. An interest in the past blossomed. Newspapers like the Goderich Signal Star and their earlier individual counterparts ran columns featuring insights into the early days in Huron. Green contributed to these insights with regular and topical musings submitted to the local press.

It is in the tradition of writers such as Arthur W. Brown of the Signal Star, Margaret Snell and W. H. Johnston of the London Free Press and later W. E. Elliot of local renown that Gavin Green wrote his early newspaper stories. His contributions to the local newspaper are noted as early as 1932. These and other articles when collated, were sufficient to warrant the publication of his first book. Publishing a book during a depression and with the looming prospect of war must have been a considerable risk. Few people could afford to purchase such luxury items. But the finished work had much local appeal. In tribute to Gavin Green and his book, a poem published in a local newspaper at the time states "Two dollars is a proper price!".

Mr. Ernest H. A. Home, a Strathroy poet and long time friend and admirer of Green, was a valued assistant in preparing the typed manuscript. Several of his poems are included in Gavin Green's books and it is likely that he acted as one of the anonymous editors who refined the mechanics and corrected the spelling of Green's rather primitive style.

In a note on the back of a photograph, Green proclaims that he closed the door on the then "Yea Old Curiosity Shop" (sic) May30, 1948. It is probable that he was open 'by chance' or at the request of a willing buyer, however, until the mid 1950s. During that decade, he set up a museum display in a second storey room of the County Library. Some of his collection of artifacts were later moved to the County Museum. The Signal Star, on July 29, 1961, published that "about 200 articles from Ye Olde Curiosity Shop ... were on display at the Huron County Museum this week." The article goes on to state that "a record attendance of over 400 persons visited the Museum last Sunday." Some of the other artifacts mentioned in the newspaper account included an old wicker-bottomed armchair used by Tiger Dunlop, Dr. Dunlop's Diary written in 1841, the diary of one David Clark, Esq., of Colborne Township, 1849, and two books recording the marriage licences issued in Goderich from 1854 to 1865.

Some of the most valued treasures in the Huron County Museum today are there only because they were collected and preserved by Gavin Green. Following his death, many of these items were donated to the Huron County Museum by Green's long time friend and beneficiary, Harry McCreath. Gavin acquired many of these treasures as the highest bidder at sales throughout the area. In this book he relates a great detective story concerning the auction of Leeburn Church, in which he traces the history of many items sold. He obviously took great trouble to learn what happened to such items connected with his past. Without his concern and appreciation for these objects, today we might not see such priceless artifacts as the Canada Company strong box in the County Museum. It is regrettable that he was not more definitive in bequesting the historically significant artifacts in his collection directly to the custody of the County Museum. Between the time of his demise and the final transfer of a number of remaining items by Harry McCreath to Herby Neill, it is feared that a number of valuable artifacts were snatched away to embellish private collections. The Tiger Dunlop diary, for example, as mentioned in the 1961 Goderich Signal Star article, has disappeared. It was never transferred to become part of the museum collection.

Six months before his death, Gavin Green donated a sum of money to the County Museum. It was used to improve the entrance to the museum and to commemorate the two major collectors, Green and Neill, whose efforts saved many priceless artifacts for future generations to enjoy. Recognizing his importance locally, Huron County Council unveiled a plaque at the museum to honour "Mr. Green, local historian and raconteur of note."

How much Gavin Hamilton Green cared for and nurtured interest in history is also seen in his donation of memorabilia and moral support for two young Colborne Township boys who had started their own "Springbank" museum in 1961.

Two small and very plain hard-covered books sit on the shelves in the County Library and in the County Archives. Should one enquire of the oldtimers in Goderich, Colborne Township and area about Gavin Green, a few are likely to proclaim with considerable pride they too have preserved an old and dusty, but highly-valued copy of one of his early books, "The Old Log School" or "The Old Log House."

Now, due to the collaborative efforts of The Huron County Historical Society and Natural Heritage Publisher, Barry Penhale, this revised and expanded edition of the Old Log School is available to the general public, more than fifty years after the original publication date. Those of us involved with this historically important publishing achievement take pride in knowing that our regional heritage is documented and available for a whole new generation of readers.

We would like to think that Gavin Green would be pleased.

This introduction was researched by Jayne Cardno for The Huron County Historical Society. Final editing and additional information was researched by Paul Carroll. Assistance in the form of an Ontario Heritage Foundation research grant is gratefully acknowledged.

Goderich, 1992.

Introduction

My Dear Reader:

As I put down these experiences of my school days in the old log and frame school houses of the sixties and seventies, flavored with other incidents that met me on the road of my boyhood journey up to eighteen years, when I bade farewell to my happy school days, I have no doubt that many who read this will say it is nonsense and foolishness, and, maybe, lies; but, my dear reader, it is all founded upon facts of actual experience. I will admit there is nonsense, and that some things are very foolish; but, nevertheless, they are true – a true picture of childhood and boyhood days of the time of which I write, the sixties and seventies of last century.

Now, I do not call a spade a shovel in my writings, but by the name that it was known by in those days. God forbid, though, that I should put in any words that might corrupt the most delicate reader's morals. However, I believe that

"A little nonsense now and then
Is relished by the wisest men,"

and so I trust there will still be a few wise or otherwise men left upon the earth to read and enjoy these writings.

But I wrote them more for the rank and file of the common people like myself, and to let the rising generation of boys and girls know how the boys and girls of the pioneer days of the sixties and seventies behaved themselves in the old County of Huron. If these records, dear reader, cause you to drop a sympathetic tear, or a tear of joy, or a tear of sweet sadness, by carrying your own memory back to your childhood and school days that will never return, I shall feel I have not written of these bygone days in vain.

GAVIN HAMILTON GREEN.
    Goderich, Ont., Canada,
    June, 1939

THE OLD LOG SCHOOL was located briefly in this log building on the same farm lot where Gavin Green was born. It was rented to the trustees of School Section No. 5, Colborne Township for $2.00 per month. It was a "free school", first opened in the school year 1856-57.

Building of the Old Log School

PERHAPS a few records of the old school would be of interest to the reader. Application to start a school was made to the Reeve of Colborne, Robert Hunt, on the 25th day of April, in the year 1856, by Sandy Annand, Peter Robertson and Thomas Grundy. According to the minutes, application was granted as follows: "This 7th day of May in the year of our Lord one thousand eight hundred and fifty-six at Goderich, T.W. Wilson, secretary, Robert Hunt, town Reeve – present, Sandy Annand, Peter Robertson and Thomas Grundy."

First meeting of school section No. 5, Colborne, was held January 14th, 1857, in the National Hotel, Colborne. Mr. Hillary Horton was called to the chair, and Mr. Andrew Linklater was requested to act as secretary.

Moved by Joseph Strong, seconded by Hillary Horton, and unanimously agreed, that the school be a free school.

Moved by Samuel Morris, seconded by Alexander Green, that James Dustow be appointed a school trustee. Carried.

Moved by Prince Morris, seconded by Joseph Strong, that Hillary Horton be appointed a trustee. Carried.

Moved by Mark Morris, seconded by James Morris, that Prince Morris be appointed a trustee. Carried.

Moved by Prince Morris, seconded by James Dustow, that Hillary Horton be appointed secretary-treasurer.

Record of a meeting held at Garbraid hotel February 21, 1857. It was decided by a majority of eleven votes to five that Peter Green's lot be the new school site in preference to Richard Cammish's and it was agreed to rent Peter Green's house for three months at a rental of $2 per month, the same to be used as a school house until the new school house be built.

BUILDING AGREEMENT

"By and between Hillary Horton and James Dustow, trustees, and James Linklater, carpenter, wherein the latter binds himself to finish the school house situated on lot 11, concession 12, Colborne, according to specifications on or before May 20, 1857, for the sum of Twenty-eight Pounds Fifteen Shillings."

The lot on which the present No. 5 school house stands, one acre in extent, was purchased from David Healey, March 12, 1870, for the sum of fifty dollars ($50.00).

AGREEMENT WITH SCHOOL TEACHER

We, the undersigned, trustees of school section No. 5 in the township of Colborne, by virtue of the authority vested in us by the fifth clause of the twelfth section of the Upper Canada School Act of 1850, have chosen Richard Haynes, who holds a second-class certificate of qualification, to be a teacher in said school section, and we do hereby contract with and employ such teacher at the rate of sixty pounds per annum from and after the day hereof, and we further bind and oblige ourselves, and our successors in office, faithfully to employ the powers with which we are legally invested by the said section of said Act to collect and pay the said teacher, during the continuance of this agreement, the sum for which we hereby become bound – the said sum to be paid to the said teacher yearly. And the said teacher hereby contracts and binds himself to teach and conduct the school, in said school section, according to the regulations provided for by the said School Act.

This agreement to continue for one year from the date hereof.

Given under our hands and seals this first day of January. One Thousand Eight Hundred and Fifty-eight.

JAMES DUSTOW,
ALFRED MORRIS,
HILLARY HORTON,
(Seal)   Trustees.
RICHARD HAYNES,
            Teacher.
Witness: THOMAS GRUNDY.

THE OLD FRAME SCHOOL was contracted for completion "on or before May 20, 1857, for the sum of Twenty-eight pounds Fifteen Shillings" to the local carpenter, James Linklater. It operated until 1949 and is shown here as it appeared around 1941.

Reunion of S. S. No. 5, Colborne
June 28th, 1935.

A REUNION picnic of the old pupils and their friends of school section No. 5, Colborne township, took place at the present school grounds today – seventy-nine years after the forming of the section. The day was pleasant, the crowd glorious. Over three hundred people – old-time pupils, present-day pupils and their friends – gathered and renewed acquaintances of old school days. I counted fifty-eight automobiles, two single buggies, two motorcycles and one truck. If some of the old pioneers, teachers and pupils, who have gone on to the land where they never grow old, could have passed by the old school, what a thrill they would have got to see the change in the process of time since they went to the old log school. Who knows but they may have passed by the old school, and viewed the scene with pleasure?

The tables were laden with dainty eatables, the children, girls and ladies were dressed in their light summer costumes, which displayed their lovely forms to perfection. I noticed some of the old No. 5 school bachelors from the Lake Shore road with fire in their eyes when they beheld these lovely female forms.

They had a fine program of speech-making, song and

A REUNION PICNIC was held for pupils and offspring of the old log school in June, 1935. Over three hundred persons attended "a fine program of speech-making song and instrumental music." The picture is from a post card souvenir to commemorate the occasion.

instrumental music, with bagpipe selections by one of the old school boys, Harold Bogie. The oldest pupil of the old log school present was John Dustow, seventy-nine years of age. Mr. Dustow gave a fine address, reminiscent of old school days, and also sang that fine old Methodist hymn, "We will never grow old over there," which he sang so sweetly, so fittingly to the occasion, that it brought sweet memories back, and a tear of remembrance dropped from many an eye.

The pupils present at this reunion who attended the old log school in the sixties were John Dustow, Joseph McCann, William Bogie (Red Bill), James and Edward Foley (twins). They were the only old pupils present that attended the old log school. They had their pictures taken, this old school gang. No doubt R. R. Sallows will have them for sale in postcard size. They would be a nice souvenir of the old No. 5 reunion to send to old friends who could not be present. Also a photo of the hundreds that attended would make a nice souvenir of the old school reunion, and 100 years hence would be interesting to posterity.

I noticed among the crowd George Sturdy from Goderich township. He went to the old log school for five years when a boy, but came to the reunion at too late an hour to get into the picture of the old school gang. He said he had a warm spot in his heart for the boys and girls of his old school days at No. 5, but was sorry to see so few of them left. He expressed his pleasure at being present, and was glad he came, though late.

Now, the present trustees, Alex. Watson, Leslie Johnston and Andrew Bogie, did the grand and should be congratulated upon the success of the old-time reunion, as they had provided swings for the children, games for the sports and seats for the ladies, elderly men and lazy pupils. They also gave everyone present a dish of ice cream and furnished free drinks – of lemonade.

FIVE OF THE OLD BOYS who attended the 1935 reunion were (L. to R.)
the twins James and Edward Foley, Joseph McCann, William "Red Bill" Bogie
and John Dustow.

CHAPTER I

In the Days of the Old Log School

I WENT to school in those good old days when A B C was plain
A B C and f-a-t-h-e-r spelt father, not "fother;" when girls and
boys went barefoot to school; when girls dressed in wincey
and print in summer and homespun dresses, ornamented with a
few rows of cloth braid, in winter. Boys wore duck trousers and
factory-cotton shirts in summer and a suit of homespun full cloth
in winter, shoes with copper toes or long boots with red tops.

Our dinners were mostly bread and treacle, with butter on
special days. Cakes, pies, apples, etc., were luxuries. A boy with
an apple could do a brisk trade at so much a bite, if he felt so
inclined. If he hogged the apple, he generally gave the core to his
chum. Boys smoked elm roots for cigars. A boy who could sport a
clay pipe smoked mullein leaves. Thus did he lord it over the elm
rooters. In those good old days we chewed each other's gum. This
was before the microbe appeared on the scene. All the "microbes"
I knew in those days were the little fellows our mothers combed
out of our heads with a fine tooth-comb every morning before we
went to school. Those she didn't get were like Mary's little lamb
and went with us to school, although it was against the rule. I

have been told that these little fellows were of Scotch descent, but I know that in those days I saw them playing tag on the heads of the English and Irish as well as the Scotch.

If there were other microbes then that tried to live inside us they had a hard row to hoe, considering the doctoring that we had to take a mixture of sulphur and black-strop molasses. Every springtime our mothers started to dope us as soon as the buds came on the tress, staying on the job until the blossoms fell off the trees. The dose was a tablespoonful every morning, with a tea-cup of salts as a chaser, or a cup of senna on going to bed. We survived, but the poor microbes must have returned to earth in a hurry, if they ever existed. However, I never saw them nor heard of them in my boyhood days.

And then they talk of your vitamins A, B, C today, when the school children have to be babied up with pasteurized milk in school hours to keep up their vitamins A, B, C. In my school days we were babied with a blue beech gad to keep up our vitamins A, B, C. I got my share of the blue beech vitamins in my school days, as I went to seven different schools and fourteen different teachers. If there were other vitamins in those days besides those of the blue beech gad, we must have got them in our eats and drinks as nature provided them, in the raw state, for I know we ate lots of raw eats, such as turnips, green apples, wild black cherries, beech leaves, sorrel, sour grass, slippery elm bark; chewed hemlock and cedar gum; drank butter-milk, skim-milk, when we could get it; drank water out of creeks, ponds, ditches, swale holes, or wherever we saw water when thirsty. Whether there were vitamins or ditamins in these eats and drinks I know not, but we flourished like a green bay tree and grew to be over six feet high.

Yes, those were the good old days of the blue beech gad and the rawhide whip. The belief was scriptural in the old log school that if you spared the rod you spoiled the child. If you were late for school you got a whipping. If you did not know your lesson

PETER AND JANET GREEN, the parents of Gavin are shown at the time of their marriage in 1860.

you got another. Whatever school rule you broke you got a whipping, and you generally got another when you reached home. In those days we took our whippings as a matter of course; but take it all in all, I think I would sooner take the whippings we got in those good old days than do the home-work of the girls and boys of these enlightened days. We had no home-work to do. Our school paraphernalia consisted of reader, copy book, bottle of ink, pen, slate, and slate pencil. They always remained in the school; there was nothing to carry home. This left us free and easy if we wished to scrap on our way.

As I cannot live my school days over again (oh, how I wish I could!) I will tell you something of the different schools, teachers and scholars of the sixties and seventies, when the sweet innocence of youth reigned in our minds.

I went to seven different schools and fourteen different teachers. I went to school until I was eighteen years of age and graduated in Third Book. The last six years of school I attended only for a few months each winter. I did not attend at all in the summer. If I write about the different schools and teachers, for you to understand my school days better, I think I should give a short account of my childhood days up to my sixth year. I feel I should mention a few happenings of my early childhood days.

I was born into the flesh at 2 p.m. on Tuesday, April 8th, 1862, in a log house in a bush in the Township of Colborne, County of Huron. I was born without a shirt; but to compensate for that I was born with a caul, or cap, on my head, so my mother told me, which may have acted as a charm to save my life on different occasions.

My first childhood battle with the elements was with fire. When I was a little over one and a half years old my mother, one day in winter, scrubbed the bedroom floor and, wishing to have it nice and dry for the night, placed a pan of hardwood coals in the room. I toddled into the room, and while going around, fell backwards into the pan of live coals. I could not get up. My cries brought my mother, but not before I was badly burnt. I have heard my parents say they had to nurse me in their arms for six weeks, night and day, most of the time walking the floor. I remember nothing of this incident, but I know I carry the scars from that pan of coals on my east end when travelling west, and will carry them until the end of this earthly journey.

The next battle for my life was at the age of two years, and it was with whiskey. It was this way. My mother had a younger brother, John Kerr, who was learning the trade of wagon-making at Benmiller. One Saturday he came to spend the week-end with his sister, my mother. As he came by Crackie's Corners (now called Loyal), where Crackie Robinson kept the tavern called the "Plough Boys' Inn," my uncle John bought a gallon jug of whiskey for twenty-five cents. It was about the cheapest fodder he could buy for a week-end treat. He landed safely at my mother's on Saturday afternoon. My mother wishing to visit at a neighbor's took David, my baby brother, along and left me in charge of my uncle John. Whether it was for pure hellery, or to see how whiskey would act on a two-year-old kid, or to keep me quiet, I never found out; but he gave me a tin cup full of whiskey. I guess I must have liked it and drunk it all, for I got a knock-out blow and lay in a stupor, dead to all the world, for forty-eight hours. On my mother's return she was frightened and thought I was dead. Just as my father came from work my mother despatched him for my grandmother and my two aunts, Charlotte and Lillie, and Mrs. McCann, who assured my mother I was not dead, but drunk. They bathed me in hot water and cold water, rubbed me and shook me, but all to no avail, as it took me forty-eight hours to sleep that jag off. So this was my first "drunk." But I do not remember anything about it. I do not think my mother ever really forgave her brother John for the pint of whiskey he gave to her first-born.

My third battle for life was with water. When I was two years old I fell into a creek. I lay on the bottom some time before I was missed by my mother. My father was mowing hay across the creek from the house. There was a board across the creek. I attempted to go across and fell in. When I was missed, mother and father ran to the creek. I was lying on my back in the creek, both my eyes open. My father got me out of the creek, caught my both feet, shook the water out of my lungs, stripped off my clothes and worked on me for some time without results. Then my father lay down upon the bed and laid me upon his chest and breathed into my lungs, and I began to show signs of life. As my father was Scotch and a Presbyterian, and was familiar with the Bible, he followed the example of Elijah when he raised the widow's son. If my father had not been familiar with 1st Kings, 17th chapter, 21st verse, I guess I would have gone to the Land where all good little boys go. Perhaps the caul, or cap, had a charm for me.

My first trip to the Town of Goderich was in a market basket, along with butter and eggs, when I was six weeks old. My mother would get up early in the morning, start off to Goderich on foot with a basket of butter and eggs on one arm and me in a basket on the other arm. It was six and a half miles from our farm to Goderich, thirteen miles the round trip, and as it was a mud road, and nearly all through bush, mother would discard her shoes and stockings and walk barefooted until she came to Dunlop's Hill, where she would put them on again to hide her bare feet from the natives of Slabtown and the aristocrats of the Canada Company in Goderich. On the return trip she would discard them at Dunlop's Hill, again feed the baby, and usually got home at noon in time to get father's dinner. I have heard her say many a trip she made to Goderich in this way, as Goderich was the only market where you could trade farm produce for groceries and other household necessities.

When I was about two years old another baby came along (my brother David) and took my place in the basket on the trips to Goderich. I was left at a neighbor's. In my third year mother left me at our own home in care of two of a neighbor's children, two little girls, one seven years and the other five years old. I suppose these two little girls thought I needed a bath, so they got the old wooden tub, filled it with water, stripped me and put me in the tub, doused me good and plenty, and kept me in the tub nearly six hours. When mother arrived home I was nearly a goner. Fortunately for me it was in July, warm weather. I guess it was fun for the two little girls, but not for me, as I have been scared of water ever since, as that was the second time I nearly lost my life by water before I was three years old and I have a faint recollection of the two girls and that old wooden tub of water. The two little girls grew up and married, but, sorry to relate, both died in their early twenties, many years ago.

As I now travel down the hill of life, I feel that I did not honor my father and my mother as I should have done, and I have many regrets when I see where I could have made their earthly journey more pleasant. No matter what I could have done I could never have repaid them for their kindness and goodness to me in the days of my youth. But as they have both passed to the Home Land, may they rest in peace. My experience is that, no matter how good you have been to the parents that brought you into this world, and cared for you from youth to manhood, after they have gone from you you will have your regrets unless you do everything you can to lighten their load when they grow old. So forget not thy father and thy mother in the days of thy youth, and when you grow old you can look back on those days with sweet recollections.

CHAPTER II

MY FATHER having learned the trade of wool carder at Logan Woollen Mills at Piper's Dam, Goderich, he moved from the bush farm in Colborne to the village of Dungannon to work in Thomas Disher's woollen mills at that place. There we lived for eight years, and that is where I spent the happy days of my childhood from the age of three to eleven. Up to my sixth year I did not know a care. Play, play, all day long, my brother David and I. There was a maiden lady of the name of Green (no relation) who kept a private school in Dungannon. She was always on the look-out for pupils, so one evening, while visiting at our home (I remember it as if it were yesterday), she said to mother, "How old is your little boy?" "Going on six years," said my mother. "You should send him to my school. What do you call him?" "Gavie. He was christened Gavin Hamilton Green." "He looks like a good little boy," said Miss Green. "Why you should make a minister out of him. Rev. Gavin Hamilton Green, D.D. Why, Mrs. Green, a name like that would look good in print. Quite dignified for you to have a preacher in the family, and with such a fine Scotch name, too."

Now, my mother must have taken in all her soft soap, for I was started to school the next week, upon my mission to be a

Presbyterian divine. The fates decreed it otherwise. When it was explained to me that I was to start to school to this nice lady next Monday morning, that night I cried myself to sleep. I thought it was something awful to go away from home, be away all day and leave my brother David all alone to play with Tom, our pet pig. We had a pet pig that grew up with us until he was four years old (pig was not pork until that age, four years), and we three played together. We used to hitch him up to a cart and ride around, and on his back. Do you blame me for crying and not wanting to leave David and the pig? We then lived in the house down from the village, where Bobby Hamilton now lives. So Monday morning, dressed in brown duck trousers and linen coat, with a penny in my pocket to pay the teacher (penny a day being the rate), I started for my first day in school. I had to come home to dinner. The building where this private school was held in those days still stands, but remodelled into a butcher shop, the last time I saw it. It was on Main Street, across the road from Anthony Black's hotel, the old Prince of Orange, which had for a sign King William on a horse. Next door to the school was a store, kept by a man named Bell. He had a son, Tommy. He also went to this private school. He is the only other boy I remember going. There were lots of girls. I remember there were three or four girls named Barr. These girls were daughters of the Rev. Barr, Presbyterian minister, who preached in the old Barr church, as it was called in those days, situated about a mile from Dungannon on the road to the Nile. There was a tollgate near the church, as this gravel road was built by a private corporation and the travelling public had to pay toll until the County took it over, and then the toll ceased. I remember the girls at school would sew and knit from 3 to 4 p.m. Now, if there are any of the old pupils living that went to this private school in those days I should like to hear from them.

I went to this private school for about a year, as near as I can remember. I learned a few things; I learned to play hookey, but did not know that it was wrong. One afternoon another boy and I thought we would play around and take in the sights before I returned to school. I guess I got tired playing; I went back to school just before school was dismissed at 4 p.m. When I came in teacher called me up to her desk and gave me a good whipping for not coming back to school instead of playing. I cried and told her, "Now you have made me cry, and I have no handkerchief to wipe my eyes." I suppose she felt sorry for me, as she took me on her knee and wiped away the tears with her apron. Then she kissed me and made me promise not to play hookey any more, and I never did. I got many a whipping from different teachers in my school days after that, but she is the only one that took me on her knee and gave me a kiss after giving me a whipping.

After this hookey and whipping, I began to learn some street talk and a few cuss words. On my return from school each day I preached these sayings and cuss words to brother David and the pig Tom; but I knew enough not to utter them in mother's hearing. David took kindly to all the sayings and cuss words, and Tom grunted his approval. This preaching all took place in the old log barn. The first thing I remember learning at this private school were rhymes that it would be indelicate to print. One was a riddle about a little dog; another was a rhyme about a monkey and a baboon. But all school children in those days knew these rhymes, although never in print. Then there was "Eenie, meenie, minie, mo, Catch a nigger by the toe, If he hollers let him go, Eenie, meenie, minie, mo." Now, this is about all this good little boy learned at the private school of the teacher who told my mother he would make a Presbyterian divine. However, it cost my father ten cents per week for tuition for me to learn a few cuss words, riddles and rhymes. My mother said it was worth the money to know where I was, even if I learned only my A, B, C and nonsense.

As I grew older I thought the preacher business looked an easy job, and I felt I should like to be a preacher. I will tell you how I got along at the preacher business later.

Time moved along, and I was taken by the hand one morning, greatly against my will, to the old log school house at Dungannon, which stood where the present brick school house stands. The boy that led me to school was a big rough boy, but kind, and he saw that I got fair play. His name was Wesley McCaig. He afterwards went as one of the Canadian Voyageurs up the Nile to relieve General Gordon at Khartoum. Wesley was wounded, and I heard he obtained a position in the London War Office. He was drowned in the Thames many years ago. I have always had a warm spot in my heart for Wesley; I think we all have sweet memories of the one that escorted us to school the first day, as it is one of the mile-posts in our journey upon the road of life that we always seem to remember.

My first teacher in this school was Andrew Forbes. I do not remember if I learned anything under his teaching, as I was in the First Book, but I do remember getting whipped several times by him. He used a rawhide riding whip on me. What the whippings were for I do not remember. My next teacher in this school was Wellington McVittie, who afterwards graduated as a Presbyterian minister. He is the first teacher I remember learning anything from; but I had always been told that I was a big dunce, so I guess the other poor teachers were not all to blame. As the old proverb runs, it is pretty hard to make a silk purse out of a sow's ear. Under his regime I also learned to tell bigger lies and use bigger swear words, and many other things that were not taught by the teacher. And one day at this school a nice little girl name Sarah, who lived up near Glen's Hill, asked me for the loan of a pin, which were scarce in those times. As luck would have it, I had one. I told her she could keep the pin, which she did. That was a new experience for me, a girl to take notice of me. Well, of course, in my mind she was my girl. But alas, this my first romance was short; it lasted about three hours. She caught me looking at her, and she made faces at me. I retaliated with an uglier face. This ended my first school romance.

MISS LEILA FEAGAN AND HER PUPILS are shown on the front steps of S.S. No. 5 Colborne in 1912. The broken batten board next to the door was noted to be still missing in 1941!

In the wintertime big, full-grown boys and girls came to school. Little boys, such as I, in those days never learned anything much in winter. We were shoved aside to make room for the big fellows. It took the teacher all his time to keep order and do the whipping. In those days the teacher always kept a boy up beside him to act as monitor while he was taking a nap or reading. The monitor's chief duty was to call out the names of the pupils misbehaving. As your name was called out you took your place in line against the wall, and as soon as the teacher awoke or finished his paper he took his rawhide and started in at the top of the line, giving from four to ten cuts, according to the gravity of the misbehavior. The boy chosen for monitor was generally one called Charlie. Charlie was no respecter of persons, and he stuttered badly. I can hear him yet calling on "Bib-Bib-Bib-Ben Mallough," "D-D-D-D-Dan Bickle,: "Tet-Tet-Tet-Tom Wiggins." Talking cost four cuts of the rawhide; playing tag around the school, six cuts; making faces at one another, eight cuts; making faces at the teacher, with your thumb to your nose, ten cuts. McVittie was a big man with a heavy moustache and side whiskers. Nearly all the pupils, big and little, were afraid of him; but one day a big boy, named Sam Pentland, from near Glen's Hill, turned on the teacher when he was going to give him a whipping. I remember the Pentland boy grabbing the teacher by the legs and tripping him. They fought rough and tumble for about ten minutes. Teacher was awarded the fight, but Pentland was a hero with all the small boys, and big boys, too, to say nothing of the girls. That was the only fight I witnessed between teacher and pupil. I have seen scraps, but that was a real Donnybrook.

One day one of the big boys, named Begley, was leaving school for good. He had it in for the teacher to even up for a hard whipping he got from him once. Now, the teacher had to go the village for dinner, and during his absence Begley gathered up all the chewing-gum (nearly every boy and girl chewed gum, about the only luxury in those days. It was pine gum, and had great chewing powers). He then got a tin cup full and put it on the stove to melt. We were all sworn not to tell. Just before the master came back to the school, Begley poured the melted gum on his chair. Master entered, took his seat, and began to call the roll. Someone started to snicker, and that set the rest going. Master reached out to grab his rawhide whip, but when he jumped up the chair stuck to the seat of his pants. He was red-hot. I see his red face yet. He had to call a big boy to help him get the chair pulled away from his pants. Then he called the whole school to stand around the walls and questioned each one to find out who put the gum on his chair. He never found out. However, he suspected that two of the big boys knew about it, so he made them take the chair outside and scrape the gum off. What gum stuck to his pants remained there. Pants were discarded for a new pair next day.

There was a small boy at school who was brought up to be very polite in his spoken words at home. Well, that was all right at home, but when he came to school and got advanced to the First Book and tackled the ox, ass and hay lesson he was greatly embarrassed. Being very polite, he balked at the word "ass" in the sentence "Can an ox or an ass eat hay?" and so got over his difficulty by substituting "backside" as a far superior term. Naturally this provided great fun for the master and pupils.

About this time there were too many pupils for McVittie to handle, so the trustees decided to put a partition through the school and engage another male teacher, which was done. I do not know whether I was promoted or turned back, but I had to go into the Junior Room under the new teacher, whose name was Sam Caesar. Brother David, although two years my junior, remained in the Senior Room. This required some explaining to my parents. Under Sam Caesar's tuition I was promoted into Part Second Book by J. R. Miller on his first official visit as inspector of West Huron schools. Previous to J. R. Miller's appointment, the

Rev. Barr (he of the four daughters at the private school aforementioned) did the school inspecting – in the Dungannon school, at least. I was proud to be promoted to the Part Second Book, which pride somewhat cooled when I found this same J. R. Miller had promoted brother David into Second Reader also, as well as patting him on the head and saying he was a bright little boy.

What boy does not remember the First Book, part two? The first picture was a big red and white bull. Most of the boys in that book thought they could improve upon the artist's model of a bull. Well, I have one of those old books at the present time. It came into my possession a few years ago. Same old bull with the same old decorations. It seems boys have not changed much in the fifty years on the bull decoration questions. I do not know how long I was in Part Second, but I knew all of it off by heart before I got promoted. I can recite most of it yet after sixty years. Such pieces as –

"Dick Fork the dunce comes late to school,
Grins, laughs and acts against the rule,
He lights a match to see it flare,
Pricks Frank's neck and pulls his hair."

And

"I had a little pony whose name was Dapple Grey,
I loaned him to a lady to ride a mile away;
She whipped him, she lashed him,
She rode him through the mire,
I will not lend my pony now for any lady's hire."

Any many more of the same tenor.

The first long boots with red tops for boys I ever saw belonged to a boy named Willie Cluff. He was a relative of teacher McVittie's wife. He lived at McVittie's and went to his school. He was a nice boy, and afterwards became the Rev.

Canon Cluff of Stratford, Ontario. He is now dead. How I longed for a pair of long boots with red tops! Now, as I belonged to the poorer class of boys, I had to be contented with the heavy cowhide shoes with copper toes. In winter I was dressed in full home-spun clothing; in summer in factory cotton shirt, blue derry pants, with braces made of the same material, whilst on my feet I wore the shoes that nature provided, with many stone bruises added on my way to school.

Along about this time the itch broke out in our school. Dr. McKay, as medical officer of health, came to school to inspect. We all had to march out into the yard, pull up our shirt sleeves to our elbows. Doctor walked past, examined our hands and arms. If we had itch we were put to one side. Well, the biggest part of the school had the itch and had to go home. I was among those who had to go home. I seemed pleased, but I did not know then the cure. Some of the aristocratic girls from the village cried when they had to come across to the itch side. That seemed to tickle us small fry, for the itch did not seem to be a respecter of persons. The cure was, stay home from school for three weeks, be washed every night in a tub of salt water. No bath tubs in those days. My, how it smarted! How David and I hollered and yelled as mother scrubbed us with the salt and water, then rubbed a stinking brown salve all over us. If I remember correctly the old skin all came off and a new grew on. However, I think all the pupils survived the itch and came back to school.

It was around now that McVittie left the school. I remember the last day – a kind of farewell. Trustees and parents visited the school. Teacher made a speech – so sorry he was leaving us. Some of the small girls started crying. We all joined in sympathy. The whole school seemed to be crying. I with the rest. Why I cried, I do not know, for I never liked McVittie.

The teacher that came to replace McVittie was named Fletcher. I still had to stay in Sam Caesar's room. I was satisfied as I did not have so many lessons to learn as my brother David, who was in Fletcher's room. Although I was poor at learning, my mother always took my part before my father. David was, of course, the scholar, but I was best to help mother around the house; so that evened things up. Along about this time we started for school one morning, brother David and I. We lived across the river, opposite Richard West's farm. When we got to Disher's hill, we always called for Ad and Jennie. Well, Ad came running out telling us that the old school was burned down last night. That seemed to be the best news I ever heard up to that time of my life, for I remember throwing up my hat and turning somersaults on the road; for, to tell the truth, I hated going to school. I played many tricks to stay home from school. I remember one morning when going to school a man named John McGrattan, a shoemaker, who had a patch of potatoes near where the Agricultural Building now stands, stopped brother David and me as asked us if we would pick potatoes. Well, as I was older than David and supposed to be boss, by right of inheritance, I said, "Yes, we will pick potatoes for you," as it was far ahead of going to school for me. We picked potatoes all day, got a good dinner and supper and a penny each. As near as I can remember this is the first money I ever earned by the sweat of my brow.

The burning of the old school did not give us boys much of a holiday. The Temperance Hall in the village was turned into a school for the higher class (Miss Blair's room), and Bickle's harness shop was turned into a school for the junior classes. I was put in the harness shop and brother David went to the Temperance Hall. A new teacher, Mr. Munro, took Sam Caesar's place. I believe Mr. Munro and Miss Blair were later married. I did not like Mr. Munro; he would whip me for not knowing my lessons. I could not learn to spell correctly. I got most of my whippings for poor spelling, but they never cured me. Poor spelling has followed me all through life, and I guess will until the end of the chapter, as I am still in Part Second as a speller along with the red and white bull.

It was at school in the old harness shop that I kissed a girl for the first time. A big girl of the name of Maggie got kissed by one of the big boys, and her cousin, named Bella (both lived in the same house), said, "Maggie, I will tell mother that Dan kissed you when I get home." Maggie thought if she could get a boy to kiss Bella it would even things up and there would be no talking. She looked the boys over and decided I looked gawky enough for the job; so she came to me and promised me a stick of candy if I would kiss Bella. Well, candy was a luxury for yours truly, so I took after Bella, ran her down and kissed her – not for my sake, nor Bella's, nor Maggie's, but for the candy. I got the candy. I met Bella forty years later, when we talked the incident over, and I hope if she is still in the flesh and sees this in print she will forgive me for telling this. Since, however, I am writing a truthful account of these school days I must tell the truth, though the heavens fall – not that I wish to show any disrespect to parties by doing so.

I remember a girl in my class used always to have a big chunk of alum in her pocket and eat it like candy; also other girls that used to eat slate pencils and chalk. I also remember an election at the time when I went to the old school that was burned down – the first election I can call to mind. Somerville was the Grit candidate and Farrow was the Tory. The Grit boys would sing,

"Somerville is a man; Farrow is a mouse;
And not fit to sit in Parliament House."

Tory boys sang vice versa. If I remember rightly Farrow was elected.

I also remember the very sad death of one of our school chums. This boy was at school Friday, and was dead Saturday. His name was Willie Stewart, and he lived near Glen's Hill. He was playing in the barn on Saturday and fell backwards off the breast beam on to the threshing floor, breaking his neck. I remember we boys of his class were very sorry, as he was a nice quiet boy at school. His funeral was the first I remember seeing. As the funeral went by the school, going to Dungannon cemetery, teacher let out the scholars and we all stood by the school until the funeral passed. I remember the pallbearers all wore flowing white streamers on their hats. The coffin was in a light wagon – no hearse then. I believe Willie Stewart was the third person buried in Dungannon cemetery. A little boy, son of Thomas Harris, was the first one, and Tommy Disher, I think, was the second. I remember little Tommy Disher was the first dead person I ever looked upon. He was in a little white casket, made by Andrew Sproul, and looked to me like marble. I remember my father taking brother David and me to see him. We did not know what it all meant, but we were very sad as mother tried to explain to us why he died.

While I am writing of deaths and the cemetery I must mention that my father made the stakes to stake out the lots in the Dungannon cemetery. He made 2,000, got a cent each, and was paid with a twenty-dollar gold piece. This was the first gold money I ever saw, and it looked just like a big new bright penny to me. Mr. Disher gave father the cedar for the stakes. He sawed them into blocks, split them with a frow, and pointed them with a drawknife on a shaving horse. David and I on our way home from school would go into the bush were father was working and pile up the stakes he made daily. The carding mill did not run all the year round, and father, being a handy man, got the job making the cemetery stakes. As he had a family of five boys by this time to feed and clothe, when he was not carding for Disher or sawing lumber for John Runciman he was making axe handles or butchering pigs for neighbors.

Along about this time my mother used to spin yarn for the neighbors. She earned many a dollar in this way that helped feed and clothe us five boys. Mother spun the yarn on the big walking wheel. I have heard it said the spinner walked seven miles while spinning a pound of yarn. Also about this time I saw my first potato bug. One Sunday my father and a neighbor named Saunby, hearing of the new potato bug, went out to our patch to see if the bug had arrived there. They found one, and brought it in on a chip with great care to show Mrs. Saunby and mother. They would not touch it with their hands, since it was supposed to be deadly poison. Father tried to kill it with tobacco smoke, but the bug just rolled on its back and played possum, same as it does today when you touch it on the vine.

Now along about this time the Saunby grist mill was built. William Givens was the contractor. This mill was built just above Disher's carding mill. Both mills are gone now. Of this once busy spot on the Nine Mile River nothing remains to tell the rising generation of the wool carded and the grain ground in the good old days. True, the old dam site is still there. But what a change from the days of the Saunbys and the Dishers, when you and I were young, Jennie.

We had now moved across the river to a big house owned by John Runciman, my father being nearer his work as sawyer in the Runciman sawmill. This house was just across the road from Richard West's farm, now owned by his grandson, Richard Park. We had a cow, and Mr. West gave us pasture for her as compensation for my helping bring his cows to and from the pasture field every night and morning, after and before going to school. A lane about half a mile long ran from the barnyard to the pasture field. They use the same old lane to this day. As I walked back that lane fifty years after I drove the cows down it, many thoughts passed through my mind, and again I saw the panorama of cows and my playmate Sarah that helped me drive them along; and I shed a silent tear, for my old playmate has passed to the land that is fairer than day. Her passing took place in 1873. This Sarah was supposed to go along with me to bring the cows. She did not always show up in the morning, and then I would have to travel back for the cows by myself. This happened many times. I thought I was imposed upon, not taking into consideration the fact that we were getting our cow pastured for my help. So I began to study deviltry; and this is what yours truly did to get even with this little girl. At this time West's farm had twelve cows, which with ours made thirteen. West's also had a "gentleman cow," that we always left in the field. The morning I put my scheme into execution I brought up the "gentleman cow" and left one of West's milk cows in the field. I remember I had a big job to get the "gentleman cow" to come and the other cow to remain behind. Well, I brought up the right number of cattle, thirteen, struck for home, and had to mind the children until mother milked our cow. Mother came back and my "mistake" was not found out until after I had gone to school. The poor girl had to go back for the other cow and was too late to go to school that day. When I came home from school at night, mother said to me, "Why did you not bring all the cows up from the pasture field this morning?" I said, "I did. I counted them and there were thirteen." "Now, do not tell me any stories," said my mother. "You brought up the John Thomas instead of one of the cows." I said, "Did I?" quite surprised. "Yes, you know very well you did," said mother, "and Sarah Jane had to go back and bring the other cow up, and was too late to go to school. Now, my boy, you just wait until you father comes home and he will tan your hide for you." Well, father came home, and mother related the incident. Humor was my father's middle name. He laughed, turned his back to me and, between giggles, told me not to do a trick like that again on a little girl. However, I had my revenge, and did not get my hide tanned either. Sarah Jane did not speak to me for some time, and when she first broke silence it was to tell me that it was a mean, mean trick.

My parents being Scotch Presbyterians, we went to the old Barr church, about a mile south of Dungannon. My father used to take David and me, and mother went when she could; but, as she generally had a baby to nurse at home those times, she seldom went. Some Sundays father did not go to church, so David and I went along, as we had been taught that little boys who went to church would go to heaven when they died, and little boys who did not go to church would go to the bad place, where there was a lake of fire and brimstone. I remember we felt very sorry for the little boys we knew that did not go to church on Sunday. So David and I were always ready to go to church on Sunday to insure escaping this burning lake of fire and brimstone. One Sunday, as David and I were on our way to Barr's church alone, on the road about where the Agricultural buildings stand we found a big horseshoe. A horseshoe was worth a penny. Two cents was some money in those days to us boys. Well, we did not want to hide the horseshoe along the fence, as someone might get it, so we decided to take it to church with us. I had on a white linen coat and I put the horseshoe under it, so no one would see it. When we got as far as the village it was getting quite heavy, and it was hard for me to keep it under my coat, so I said to David, "Let us go to the English church," which was in the village. He agreed, so to the English church we went, horseshoe and all, and I sat through the service with the horseshoe hidden under my coat. We then carried the horseshoe home and hid it in the henhouse, took it with us when we went to school Monday morning, and sold it to George Videan, the blacksmith, for two cents, with which we bought two sticks of candy at Johnnie Roberts' store and had a treat.

We said nothing about the horseshoe or the English church business to our parents, but David and I decided that in the future we would attend the English church. We liked to see the preacher in his white gown in the low pulpit and in his black gown in the high pulpit. We liked the singing. People smiled and looked pleasant at one another, while in Mr. Barr's church people always looked so sour and sad. We were always afraid to look around in Barr's church. I might here mention some of the choir of the English church in those days. I remember William Holland, a big, fine-looking young man with black curly hair. Also Mrs. Burritt and her four daughters sang in the choir then. These Burritts lived at the Nile. Well, David and I decided to go to heaven with Mr.

Jones' church instead of Mr. Barr's church. We were getting along splendidly. Had the litany of the service nearly all off by heart. I remember to this day that I then decided to become an English Church preacher. But, alas, the horseshoe that I carried to that church on my first Sunday there, its luck failed us in this way. A neighbor woman visiting at our place said to mother, "I saw your two little boys at our church on Sunday." Well, mother, being a stiff Presbyterian, soon stopped that performance, and back to Barr church we had to go. We put in a protest. I told mother that if we could not go to Mr. Jones' church we would not go to any; that we would go to hell with the other boys that did not go to church. Somehow or other we had found that "hell" was the other name of the bad place that burned with fire and brimstone. Mother wanted to know where I learned the word "hell." I told mother I had heard Mr. Barr, the minister, say "hell" in church. But that did not save me. I got a good whipping from mother for saying the word "hell" instead of "the bad place where bad boys went," and was told not to say that word again. David and I felt very badly, as we knew Mr. Jones, since he often went fishing and hunting with our father. He was a jolly fellow, would laugh, smoke, play cards, etc.

I still had it in my mind to be an English Church preacher, and occasionally David and I continued to rob the old Barr church of two of its congregation by going to Mr. Jones' church. Then I began to preach every Saturday afternoon! Saturday about, we boys and the Saunby and Disher and West children visited one another at our respective homes to play. These visits generally ended in a fight, the visitors calling the hosts nasty names until they were out of hearing. This all seemed to come as a natural course of events as we were all good friends again first time we met. When the Saunbys and the Dishers came to our place to play I always preached. I preached in the stable, using boxes for pulpits, mother's white nightgown for the first part of the service and a black silk shawl of mother's for the high pulpit part. I must have been very eloquent, as the congregation always remained and listened to my discourse. I recited the litany part, and what I preached I guess were parts of sermons I had heard at both the English and Presbyterian churches. My sermons were a cross between the two, flavored strongly with hell fire and brimstone. After I found out that the bad place that mother said bad little boys went to was called "hell," and that there was a big lake of fire and brimstone, I seemed to like to dwell on the word "hell." I seemed to like to say it better than "the bad place." It was shorter and had the punch. My congregation, too, liked to hear me say "hell fire and brimstone." I told them that if they were bad boys and did not say their prayers at night, told lies, said swear words, or made faces at the schoolmaster when he wasn't looking, that was where they would all go; or if they whistled on Sunday or cut a stick with a jack-knife, or caught grasshoppers or lady bugs on Sunday instead of going to church, they would go to hell sure. I seemed to get so much hell fire and brimstone in my sermons that I was frightened of my own preaching. I know I always ended the preaching with these words, "Let your light shine before all men," to let the congregation know to get their collection ready. I took up the offering myself, which generally consisted of pieces of slate pencils, buttons, bits of candy , pieces of colored glass, sometimes a bit of cake. Once I got a copper. Out of my old congregation I think there are only three living, and they have so far escaped the burning lake of fire and brimstone. These are Adrian Disher, George Saunby and my brother David. (P.S. – Since the above was penned, George Saunby has passed on to the other world. I hope he escapes that lake of fire and brimstone that I preached to him in the days of my youth. Now, only two remain of my old congregation, Adrian Disher, brother David and myself. And as we are all past the threescore years and ten, and then some, we must be nearing the lake of fire and brimstone ourselves, which I trust we also will escape when we reach our earth's journey's end.)

Along about this time the first Methodist church was built in Dungannon, and as the Wests, Saunbys and Dishers were all of the Methodist persuasion, and as we all played together and went to day school together, and as there was a Sunday school started in the new church, my mother thought it would be all right for David and me to go to the Methodist Sunday school. I remember the first superintendent was Richard Treleaven, sr. My first Sunday school teacher was Miss Susan Anderson. Afterwards she became Mrs. Hugh Girvin. Later my Sunday school teacher was Robert Treleaven. About now one of our day school and Sunday school playmates died suddenly. Her name was Sarah Jane West – the little girl who helped me drive home the cows. She died April 6th, 1873, in her early 'teens. David and I took this death very sorrowfully, as we were just beginning to realize what death meant. We lived just across the road from her home. The death of Sarah Jane West cast a gloom over the pupils of the day school and the Sunday school as well, and they attended her funeral in a body. She was handsome and loved by all.

I remember going home from school one night with brother David and the Disher and Saunby boys. We met two big boys from the village (Ben Crawford and Bill Bickle), who had been down at the river fishing. When we had got past them what we thought was a safe distance, we started calling them names. They then threw stones at us. We dodged the stones as we walked backwards, but it happened that there was a horse and buckboard coming from the direction of Port Albert with the driver asleep or drunk on the seat. The horse was right on me before I knew it, and when I turned around he trod upon my big toe. We were about half a mile from home, but with David's help I got there. Then I had to tell mother a story, for if she had known we were calling anyone names she would have whipped me. But mother soon found out how it happened. She told me it was my punishment for being a bad boy. I guess she was right, for I suffered enough punishment with the toe before it healed and a new nail had grown on. And such a nail as the new one was! I still carry it as a trade-mark of my school days, and as I look back I still see that old white horse and buckboard and the big man with the bushy whiskers asleep on the seat.

But losing my big toe-nail did not cure me of calling names. David and I had a little red dog with a short tail. Well, we three pups grew up together. Dog's name was Tuppence – Tup for short. He would not let anyone touch us, not even father. David and I would go out to the road on Saturday, hide in a fence corner, and call people names as they passed by on foot. If they stopped to say anything to us, we would say, "Sig him, Tup!" Well, one evening a big boy named Hagen, who lived down the 4th concession was going up to Dungannon. We saw him coming, and got out to the fence. When he came within hearing distance, we started calling him names. He started pelting us with stones. We said, "Sig him, Tup!" so he sigged him and bit him. The Hagen boy complained to my father, and next day father thought he must get rid of the dog. So it was that when David and I came home from school no dog met us. Poor Tup was gone. That day Andrew Green had been up to the carding mill with wool, and father had given him the dog. David and I put up such a howling match of crying for the dog that we had to be kept home from school, and when Sunday came father had to get a neighbor's horse and the light wagon and go eight miles for the dog. David and I had to go along, as the dog would not follow father. If the dog had not been tied up he would have been home post-haste. You should have seen us three pups welcome one another. Poor Tup got killed shortly after; a neighbor woman stabbed him with a pitchfork. David and I brought him home on a hand-sleigh and buried him. We cried for three days and had to stay home from school again. If anyone mentioned Tup we cried for many days afterwards. We shed many a tear for Tup. Father got us a little black and tan dog, but I never took to him, nor have I taken to any other dog to this day. Have heard it said there are dogs in the other world; if so, I hope to meet the little red dog Tup of our childhood days, when we were pups together and spent many happy days in one another's company, David, Tup and I.

The people who lived across the road from us, the Wests, had a peacock. He used to get on the barn and cry before rain. I used to get on our house and mock him. Now, it had been very dry weather for some time and the old peacock never cried for rain. One Sunday evening father and mother went down to the mill at the river for a walk, taking my brothers John and Andrew along, as they were small, and leaving David and me at home. Father said to me, "Now, Gav., don't go on the roof of the house," as I used to go on the house and run

YOUNGER BROTHER DAVID and Gavin were very close friends. From their pre-school days where they lived in an old log house at this spot on the edge of Dungannon until his death in 1958, David and Gavin maintained strong fraternal bonds.

along the ridgeboard. Well, West's old peacock flew on to the barn and gave several cries and stopped, so, as mother had no rain-water, I got up on the house to start when the peacock left off. I felt it would be sure to fetch rain in peacock or I kept on crying. I could imitate the peacock pretty well. I thought the good Lord would not know the difference and send the rain. Then I spied father and mother coming up the gully from the mill. Of course they saw me on the roof of the house, and heard me also. Down I got and went to meet them as if nothing I had done was wrong. I saw father stop and cut a switch, so I knew I would get a whipping for disobeying orders by going on the roof of the house and breaking the Sabbath; but when father grabbed me to whip me, the dog, Tup, flew at father and grabbed him by the throat and hung on to his coat collar until he threw the switch away and mother made him let go of father. Thus he saved me from a whipping for breaking the Sabbath by going on the roof of the house, so why shouldn't I cry for that little red dog when he got killed? Well, the peacock and I between us brought rain that night; mother got her rain barrel filled, and we saved the potato crop also. I should have put the peacock episode in before the little red dog was killed, but it did not come to my mind until I got this far. It all happened, however, before we left Dungannon, so I must put it in before we move, which we did shortly. I might here add a little more about the peacock. This bird would fly up on the barn and cry, and as I had no barn I got up on the house roof to mock him, which I could do very well, for he always answered me when I gave him a cry. We were always good friends. When I went over to West's he would strut around and put out his feathers for me to admire. I used to strut around, too, but I had no tail feathers to show off with.

AT THE OLD WEST FARM, Gavin visited with the great grandson of pioneer farmer Richard West to reminisce about the days of his work as a chore boy and to recall certain escapades at the farm near Dungannon at the crest of Disher Hill, West of the village.

One more incident, before leaving Dungannon, shows the innocence of youth in those days. One night a neighbor man called and woke us up, saying, "Pete, there is war in the camp." David and I slept upstairs. We jumped out of bed, looked down through the crack in the board ceiling and saw mother going away with the man to the war. David and I cried ourselves to sleep; we thought mother had gone to the war and we would never see her again. We could not make out why father did not go to the war instead of mother. If father did get killed at the war we could get along without him, but not without mother. But when morning came and we got up, mother was home from the war and getting breakfast. David and I were glad, and we wanted to know all about the war in the camp. All she told us was, "Mrs. Saunby has a young daughter." We could not see what that had to do with war in the camp, nor did we know until a long time afterwards.

CHAPTER III

IN THE YEAR 1873 we bade farewell to Dungannon, where David and I had spent many happy days of our childhood. My father moved to Port Albert to work in Mr. James Crawford's mill. I was eleven years old when we landed in Port Albert, and David was nine. In those days Port Albert was quite a flourishing village. Boats came into the harbor for cordwood and tanbark. There was a grist-mill, and a saw-mill, a barrel stave and heading factory, a shingle-mill, a blacksmith shop, and two hotels, one kept by Ben Wilson, the other by George Graham. Besides these there were two stores. One was kept by Thomas Hawkins, who also had the postoffice and the telegraph office. Mr. Crawford kept the other store. Port Albert is where I did my first regular manual labor for hire. David and I used to carry out and pile the barrel staves from the cutting machine. We also piled the barrel heads as they fell in pieces from the heading machine. The stave bolts were put into six steam boxes holding two cords each and steamed for twenty-four hours to soften them. A man with a hook pulled them out, loaded them on a wheelbarrow and took them to the stave knife, where James Crawford, jr., cut them into staves. David and I carried away the staves and piled them in the yard for two York shillings per day each. We had to wear coats and mitts, as the staves were hot and would burn our arms if not covered. One day it was cut staves, the next day cut headings. Staves and headings were hauled to Goderich by team, where they were made into salt barrels by the coopers.

When not working, David and I attended school. The year we were in Port Albert, 1873, the new school-house was built. It still stands at the present time. The school teacher was a young man from Goderich, John Dickson by name, son of the jailer Dickson. He afterwards took up medicine and rounded out his career as a practising physician at Portland, Oregon. I remember getting a whipping from this teacher with a rawhide riding whip. My both hands were blistered so that I could hardly shut them. I thought I did not deserve this whipping, and swore vengeance against him when I grew up, but my wrath cooled in a few days. There was a boy in school one noon hour called another boy a "d——b——." Of course this boy told the teacher, and the other boy was called on the carpet. Says the teacher, "Joe, did you call Tom that bad name?" "No, sir." "Well, what did you say?" "I just said to Tom, 'McCrea has a dandy buggy.'" With this so well turned, the teacher smiled and told Joe and Tom to take their seats. We had all expected to see Joe get a good thrashing, but I guess the teacher was in a better humor than when he whipped me for less. I left this school and teacher without a tear, but there were some nice boys and girls who went to Port Albert school. Some of them I have as my very dear friends to this day. I remember some very pretty girls also went to Port Albert – Kate and Lizzie Hawkins, Grace and Lizzie Crawford, Annie Graham, Maria Young, Sarah and Mary Dunbar, and many others.

There was no church at Port Albert in those days, so I drifted away from the preacher business. The world seemed to be getting larger. I found out I would have to be a scholar before I could be a preacher. All I knew was that I was not cut out for a scholar and a gentleman, so my ambition for any easy job at the preacher business vanished. But still I had an ambition for an easy job. So I thought the next easy job was a storekeeper's. I could see lots of nuts, candies and raisins, cheese and crackers, red herrings, etc., to eat while on the job. Well, the storekeeping business eventually materialized about forty years afterwards. There did not happen to be any eats in the store, but the easy job came in all right. So one ambition of my youth was realized.

I remember mother had a hard time to make us boys keep the Sabbath holy at Port Albert with no church nor Sunday school to go to. I also remember a boy going in swimming and getting drowned. His name was George Dreaney, and he was the son of the village blacksmith. This sad event cast a gloom over the school children and village. There were some wild men passed through the village in those days. They were real giants. They came from Kingsbridge and Kintail. I remember two of these big Ashfield boys, big Jim McAdam and big Jack Pierce. Jack had the biggest moustache I ever saw on a man. It was said he could tie the ends together behind his head. When these boys stopped at the Port to liquor up on their way home from Goderich, the natives generally gave them a wide berth. I remember big Jack Pierce cleaning out George Graham's bar-room one night. He chased the natives and George Graham out, then commenced to clean house. He pitched everything out of the bar on to the street. He then grabbed the big box stove with a fire in it and pitched it out on to the street also. Then he started for Kintail. It was customary in those days for the big fellows to clean out the bar-rooms of the country taverns when they had got full of whiskey in Goderich. They were like a steam boiler carrying 100 lbs. of steam without a safety valve. That was when the country tavern bar-room had to suffer, for there it was the big fellows let off their steam. But I never heard tell of these Ashfield boys cleaning out the bar-room of big Anthony Allen's tavern at Dunlop. Oh, no! They knew they would get measure for measure from Anthony.

CHAPTER IV

AFTER we had been living about a year in Port Albert they wanted a head sawyer in the Stanley Mills at Goderich. As they paid higher wages than Mr. Crawford, we moved to Goderich. David and I went to the old Central School. I was in the Second Book, but in the junior room. Miss Longworth was my teacher. David had passed me by and was put in the senior room. I think his teacher was Miss Norval. W. R. Miller was principal. I remember Miss Longworth taking me down to Mr. Miller's room to get a whipping. I forget now what it was for. I do not think I learned much in the Goderich school. I did not like the town boys nor the school. I remember one Saturday David and I were going fishing. Passing the stone house on East street where Capt. Thos. Dancey then lived, a big Dancey boy came out and took our fishing pole from us, and we never saw it afterwards. Well, I swore vengeance against this big Dancey boy, but he turned out to be a husky fellow and a lawyer, so I gave him the benefit of the doubt and licked thumbs.

When we lived in Goderich (1874) Lord Dufferin visited the town. He was entertained by M. C. Cameron. There was a welcome given to him at the Central School. All the school children of the town assembled in lines on the lawn of the school. We sang a song of welcome, composed by Inspector J. R. Miller for the occasion, and an address of welcome was read by Johnnie Robertson, elder brother of the present editor of The Signal-Star. I might also mention that I am an elder brother and that when there is anything I am not clear on about our school days, all I have to do to get put right is ask my younger brother, David. So younger brothers are not all born in vain. I remember I was very much disappointed in Lord Dufferin. He just looked like an ordinary man with a short chin whisker, straight hair and long nose. Well, that is how he looked to me. I liked his aide better, with his nice red coat and sword hanging by his side. I thought sure he was the Governor-General. But when father pointed out the real lord, I thought my father was a better-looking man than he. The old town was decorated for the occasion. One feature was typical of the town's then chief industry, that was a salt barrel arch erected on West street, under which the Governor-General and suite passed on their way from the harbor up town. There was a society ball held at night in the old Agricultural building, or skating rink, at Victoria Park. This building was long since demolished, and is as much a thing of the past as the gaiety of the occasion.

When we had been in Goderich a little over a year the Stanley sawmill burned down. The Goderich fire-engine, then new, was used at this fire for the first time. This mill stood where the old Wheel Rigs building stands today. With the burning of the saw-mill, father lost his job. Robert Johnston, at Sheppardton, wanted a sawyer for his mill and rake factory, so out to Sheppardton we moved, where my father and mother both lived until the end of their earthly journey. When we landed in Sheppardton, David and I were both in the Second Book. I remember some of the lessons yet, such as "Billie and Nannie, or The Two Goats;" "Silver Locks and the Three Bears;" "Little Red Riding Hood;" "Tommy and the Crow;" "Little Bo-Peep;" "Old Mother Hubbard;" "My Father's at the Helm;" and "The Wind in a Frolic," which ran:

"The wind in a frolic sprang up from its sleep
Saying, "Now for a frolic, now for a leap,
Now for a madcap's galloping chase,
I'll make a commotion in every place."

I also remember "Love One Another" –

"A little girl with a happy look
Sat slowly reading a ponderous book
All bound with silver and edged with gold,
Its weight was more than a child could hold."

My first teacher at Sheppardton was a lady named Miss Dobbie.

The first winter we were at Sheppardton was a very hard one, and as there were five boys in our family, besides father and mother, to feed and clothe, and father as head sawyer in the mill got only $1.25 per day, sometimes we were on short rations. Well, there was a farmer near where the old St. Andrew's United Church now stands, not far from Port Albert. This farmer had no boys and wanted a school boy to do the chores night and morning for his board. This was in the winter. As I was the oldest I was packed off to this farmer, and so back to the old Port Albert school I had to go. I had lots of chores to do, as the boss and his hired man teamed wood to Goderich every day. I was well fed, but lonesome. The daughter, younger than I, was a nice little doll of a girl, but she would not play with her father's chore boy; so I had to play by myself when I got my work done. I was at this place about six weeks, when I got fired. It happened in this way. One morning when I came up for breakfast after doing the chores, the mistress said to me, "Take that grease and grease Amelia's shoes." I said, "I will not." "You won't?" said she. "No," said I. "Well, then get your breakfast and go home." The word "home" sounded good to me, for be it ever so humble, there is no place like home. I was glad to go home, as I always got a welcome from mother. On entering the house, mother said, "Why, Gav., you home. Are you sick?" "No, mother; Mrs. Scott sent me home."

"What for?" said mother. "Because I would not grease Millie's shoes." "Why did you not grease them and do as you were told?" I told mother I was not going to grease a girl's shoes. Mother said, "You big goose; I have a notion to send you right back." But she didn't. I often see Millie. She has been a grandmother for years, but she is a dandy little lady yet, and often as she passes me by I think of the morning I got fired for not greasing her shoes.

My "Home, Sweet Home" coming from Mrs. Scott's was short-lived. I was soon lent out again for my bed and board and schooling. John Quaid, who lived near old Port Albert Presbyterian Church, having heard I was sent home from Mrs. Scott's for not doing as I was told, he, having no children of his own, thought I might be a good boy to do the chores. So here I was sent out again and back to the old Port Albert school. The boys and girls I liked, but that man teacher! If I had been big enough I could have chewed him up without salt. Chores at Mr. Quaid's were hard work. I had to pull straw out of a stack with a wooden crotch stick, like a harpoon or fish hook, only a straight handle. Easy to shove stick into stack, but manual labor to get it out again with a bundle of straw attached to it. About two hours' work each night after school to get enough straw to fed five head of cattle. I had a much harder job than at Mr. Scott's. I often wished I had greased Amelia's shoes and stayed at Mr. Scott's. I was well fed at Quaid's, but the bed part! Nice bed, but my bed-fellow, the hired man, a big husky Scotchman named Ronald, with a big black bushy whisker, and when he undressed for bed I was scared of him, as he had so much hair on his body. No nightshirts nor pajamas in those days for hired man and chore boy. Well, I sashayed away from Ronald as far as the bed clothes would allow. But the snore I got from Ronald kept me from having the sweet slumber that I should have had to go with my bed and board. I asked Mrs. Quaid to let me sleep with Mary Jane instead of Ronald, but for some reason unknown to me she said no, that I must either sleep with Ronald or go home. So I stayed on the job and slept with Ronald. I often wished that I had been born a girl instead of a boy, so I could have worn a big long nightgown down to my feet when I had to sleep with Ronald.

At. Mr. Scott's I could play by myself when I got my chores done. Had no time to do this at Mr. Quaid's. About the only amusement I had, not real fun for me, was watching Ronald trying to make love to Mrs. Quaid's niece, Mary Jane. But as all things upon earth come to an end some time, so when the spring came around and the grass began to grow and the straw all pulled out of the stack, I was sent to my home again. As Mrs. Quaid was a Courtney from Nova Scotia and a Presbyterian of the old school, I

"MY HAIRY BEDFELLOW, a big husky Scotchman" named Ronald Good sported "a big black bushy whisker." Much to Gavin's dismay, chore boy and hired man had to share the same bed at John Quaid's farm.

was sent home with a Scotch Presbyterian blessing – "For not to play marbles, whistle or laugh on Sunday, and if I must sing, let my singing be the Psalms o' David, and to go to church and Sunday school, do as I was told and be a good boy, and I would go to Heaven when I died." But I got so much Shorter Catechism with this straw pulling that I thought being good and hard work all went together; so when I got to home, sweet home again I began to kick both legs over the traces and I still have one leg over the traces yet, after all the good advice I got from this good and kind lady, Mrs. John Courtney Quaid.

P.S. – After my younger brother David (the brains of the family) reading this paragraph, he tells me I forgot to mention about the steer Mrs. Quaid gave me. I would have felt very sorry not to have mentioned Mr. and Mrs. John Quaid for their kindness in the fall of the same year. Mr. John Quaid came down to our home at Morrish's Mill leading a two-year-old steer as a present to Gavin for staying on the job, pulling the straw out of the stack to feed the cattle, and doing what Mrs. Quaid told me to do and being a good boy and not saying bad swear words to the cattle or driving them to water with a pitchfork. Well, I strutted around the mill and down the boundary line to Sheppardton with my chest out trying to sell my steer. I felt as big and important as Big Bill McLean, the cattle buyer of Goderich. But no one would buy my steer, so we fattened steer on turnips and killed it for beef. Weighed 400 dressed. I got my share of the steer in eats and the hide to sell for myself. So getting fired at Mr. Scott's for not greasing Amelia's shoes, I got a good character recommendation and a two-year-old steer. Whether it was fate, or fulfillment of the Scriptures, "Cast thy bread upon the waters, and it will return to thee after many days," buttered – this I know: I was some drover and the proud possessor of a steer for a couple of months. The next fall after the shoe greasing at Scott's, and the straw pulling at Quaid's, I was loaned again for my bed and board. I had an uncle living in Tiverton, who was engaged in the butchering business. He was mother's brother and had no boys, only girls, at that time. He said to mother, "Jennie, let me take Gavie home with me. I will send him to school." One mouth less to feed; so I was packed off with my uncle to another home and school among the Highland Scotch of Bruce county. I expected to have an easy time at my uncle's, as he was a jolly fellow, but I found out I had a lot of work to do after school hours – carry in wood, clean the fat off the guts at the slaughter-house, and keep fires on Saturday to render fat in tallow. I did not like the butcher business. To see the cattle slaughtered turned me against meat. Neither did I like this school. The windows were about six feet from the floor. We could not see out and seemed to be closed in from the rest of the world. A strange boy going to a new school generally has a hard time, for he is a mark for the bully of the school. At this school there was a big boy, named C——, who used to call me all kinds of nasty names relating to the butcher trade. Well, I either had to fight him or eat crow. To tell the truth, I was afraid of him. He was a husky boy and bigger than I was. But I had to challenge him to fight, so I had the customary chip placed on my shoulder and started to walk around the school-yard three times, daring the C—— boy to knock it off. If he did not knock the chip off on any one of the three rounds he was considered licked. He did not knock the chip off, and no one was gladder than yours truly, for I was pretty sure he could have trimmed me. C—— and I never licked thumbs, as was the custom when two boys quarrelled and wished to make up friendship. (Each licked his thumb, then touched thumbs, and friendship was renewed.) After this episode I had a better time at school. I was considered a kind of hero; but honors came to me only by a bluff. I remember one girl, Marion Patterson, who always took my part from the first day I went to that school. I did not learn much there, as I promoted myself back to Junior Second, where I always knew my lessons and had a easy time.

My uncle was married to a Highland Scotch girl named Christine Morrison. She had a sister named Bella Morrison and a cousin named Annie McLean. They were both about twenty years of age. They made fish nets. My auntie also had a brother named Angus, a wagon-maker. These people all boarded at my uncle's, and they all talked Gaelic most of the time. Well, they had their fun. These girls would teach me Gaelic words and sentences to say to my uncle and auntie and Angus; then Angus would teach me Gaelic to say to the girls. I thought it great fun, but did not know what it was in English that I was told to say. I remember Bella Morrison chasing me out of the house with a broom after saying some Gaelic to her that Angus had told me to say.

The year I went to school in Tiverton was the year Mrs. Turner's big brick hotel was built. I went to church in Tiverton, but did not like the Gaelic church where my auntie went, so I patronized the Baptist church. They had a big stone fountain in church to baptize the converts in. I liked to see them getting baptized; it was a novel experience to me. I thought it a great sight. I determined I would be a Baptist.

Now I must tell you of an amusing incident that happened to me at my uncle's when in Tiverton. My uncle had a boy and a girl friend named Sammy Fisher and Fanny Bissett living near Benmiller, in Colborne township. They got married and came to my uncle's on their honeymoon. As it was in the wintertime, my uncle's house and beds were all filled and they had to change around to get sleeping quarters for the bride and groom. Finally things were fixed up, but I was left out. No place to sleep and no bedclothes. Then it was arranged I should sleep with the bride and groom. The bride and groom sized me up, and I guess they thought I looked green enough to sleep with them. I was put in the back of the bed, but I do not remember whether it was the bride or groom slept next to me.

I hardly know what to call it, an honor or a disgrace, the way those two big girls, Bella and Annie, teased me about sleeping with the bride. I could not see where the fun came in. But I bunked with the bride and groom for a whole week when they were on their honeymoon. About forty years after this incident I called at a home in Goderich, and the lady of the house introduced me to her sister. "Mrs. Fisher, this is Mr. Green. I guess you never met him." "Oh, yes," I said; "I used to sleep with her." "Oh, good Lord! What do you mean? Who are you, anyway?" said Fanny, quite shocked. I said, "Tiverton." Said she, "Are you the little wretch of a boy that slept with Sam and me at Johnnie Kerr's at Tiverton?" "Yes," I said, "I am that little boy." We shook hands again and had a good laugh over the incident. As you read this, I hope you have the same. Fanny was a childless widow at the time of this meeting and has since passed to the other world. I trust to meet them both again in the sweet bye and bye. I mention this incident of my boyhood school days with all due respect to their memory.

When spring came around I was shipped back to my home at Sheppardton. I went to work in Bob Johnston's rake factory, shaping one end of a rake tooth with a machine. Brother David shaped the other end of the tooth in another machine. These teeth in the raw were about seven-eighths of an inch square and six or seven inches long. To make the tenon end we held the tooth with iron tongs with flat jaws and pushed the wood in a machine with revolving knives. To make the rake end of the tooth a similar process was gone through, only we used tongs with a concave surface. I forget how many teeth constituted a day's work, but it was plenty. Our remuneration was fifty cents per day each and chew your own. This job lasted about a month; then back to school we went until the holidays. Then to the berry patch to pick berries to sell in Goderich. I remember my brother David and I starting off one morning on foot from Sheppardton for Goderich with a pail of raspberries to sell. We carried the pail between us on a stick, each carrying one end of the stick, with the pail in the centre. Well, we walked the eight miles to Goderich and then walked around the town looking for a buyer. We spied a house whose owner we thought by the look of it might buy our pail of berries. In we went, and a very nice lady came to the door. She asked us our names and where we came from, etc. I shall never forget her, for she was so nice and kind to us. She paid us fifty cents (all we asked) for our pail of berries and gave us all we could eat of cake, pie, and other good things. I found out afterwards this was John Elwood's house. We afterwards sold berries in Goderich, but this was the finest reception we ever got. Elwood's house is still standing, and as I pass it I remember the good feed I got in it when I was a hungry boy. Mr. Warnock, the painter, lives in this old house at present.

A few more days of berry picking when an old lady came to my mother to see if she could borrow the loan of "Govy" for a couple of months. She would give me my board and washing, and if I was a good little boy a present when I went home. Well, board was eats in those pioneer days, and I knew I would be well fed – which was something my belly had longed for in the last few years. So I was engaged. I found I was to act as a kind of companion, flunky and private secretary, without any writing, authority or pay. My duties consisted of feeding the hens, pigs, cows and horse, and milking the cows, straining the milk, churning the butter and driving the old lady to Goderich town every Saturday with the butter and eggs, which were traded for groceries, tobacco and a jug of whiskey. We traded at Geo. Grant's grocery, on the Square – it was on the site of the present Maple Leaf bakery. All grocers sold whiskey in those good old pioneer days. The old lady always had her toddy before going to bed. Visitors all got a horn of whiskey if they wished. Not many refused. The old lady smoked a short clay pipe. This is where my private secretaryship came in. I cut the tobacco, filled the pipe, lit it and got it going good, then handed it to her ladyship.

When we travelled to town the old lady always smoked her pipe on the road. She sat back in the old phaeton and enjoyed her old clay pipe. If we happened to meet anyone on the road she handed to pipe to me to keep the fireworks going until they passed by; then I handed the pipe to her again. When we reached Slabtown (Saltford was Slabtown in those days) I had to look after the pipe of peace until we came to the same place on the return journey. The old lady did not smoke while in town; but once through Slabtown on the return journey I lit up the old clay pipe and handed it to her and she kept it going until we landed home.

Well, I had to milk two cows, night and morning, strain the milk into big flat earthenware milk pans. They were the kind used before the tin pans came on the market. I had a little quart pannikin that I had to save the strippings in (about a pint of the last milkings from each cow), which was richer. This was saved for our porridge and the old lady's tea. I got no tea. My drink at meals was skim-milk, often sour, but I evened up by milking about a pint into my little pannikin and drinking it before I stripped the cows. One morning I forgot to take my drink of new milk, so I drank the strippings, and refilled the little pannikin with new milk out of the pail. I found out that the strippings were so much richer and creamier than the ordinary new milk that I drank the strippings ever after and gave the old lady the little pannikin of new milk for our porridge and her tea. Any that was left over we put in the churning cream crock. I was waxing fat on the strippings, when one morning the old lady said to me, "Govy" (being Scotch, she put the soapy side to my name, which was Gavin), "do you feed the cows and water them regularly, as there is something wrong with the strippings, they are not any richer than ordinary milk." I began to shiver and shake and said: "I guess I have forgotten to always feed them their chop."

Well, the scare gave the old lady her strippings again and I had to get along on ordinary milk. The strippings had spoiled me for drinking ordinary milk, and I now tried some of the cream in the cream crock. It was too thick to drink and the old lady always skimmed the cream off the pans each morning. I could not drink any of the cream off the pans without her noticing they had been tampered with. However, "necessity is the mother of invention," so I got a straw and tried the milk pans. Morning's milk would just be about right for skimming the cream off at night with my straw. It left no mark on the cream; you would never know it had been robbed. About twenty-four hours after this when the old lady would go to cream, the pan would just give a little skim of cream off the pan I straw-robbed. So the old lady said to me, "Govy, take two pails and keep each cow's milk separate, so we can find out which cow is failing in her milk. Very little cream on some of the pans." I said, "All right."

I divided my custom as fairly as I could between the two cows; one night I robbed one cow's milk, next morning the other's. One morning the old lady says to me, "Govy, I believe our cows are bewitched; some old witch is stealing our cows' cream," and as there was talk in the neighborhood of cows, cream, crocks and churns being bewitched – well, I thought, that was all right for me, and I kept right on in the creamery business stronger than ever. So one night the old lady said to me, "Govy, you hitch up the old horse in the phaeton tomorrow morning and we will go to town and see Doctor Ure, the Presbyterian minister, and fetch him out, as I know our cows are bewitched, and see if he can find out who the old witch is that is stealing our cream."

They say everything comes to him who waits; that came to me in the shape of jails and preachers. The big Goderich jail came up before me in the night as I lay awake. My bedroom seemed to be filled with jails, preachers, cows, milk, cream and strippings. Necessity the mother of invention came to my aid again in the shape of a big white lie – to tell the old lady in the morning that I had forgotten to feed the cows their chop for a long time, and I guessed that was the reason there was so little cream on some of the milk pans. The old lady seemed to be pleased to hear me say I had forgotten to feed the cows their chop, for I believe she really thought her cows were bewitched. I had expected to be sent home for not feeding the cows their chop, but the cream came back and I was kept on. I was afraid of Doctor Ure, the preacher, coming out and questioning me about the cows. Hence the white lie, as he would have found the witch that stole the cream.

I was afraid of preachers, as I was taught in my home to have great reverence for preachers, as they were so good that they knew everything, and I was afraid if Dr. Ure came out he would know I had stolen the cream and would send me to jail. I was more afraid of the preacher in Goderich than of the good Lord in Heaven, for I knew He would forgive me for telling the lie, as I was only a little boy, and always said my prayers at night, and sometimes in the morning, went to church and Sunday school, and always put my copper on the collection plate, never whistled, whittled with a knife, nor killed chipmunks on Sunday.

After losing the strippings and cream, I turned to the old lady's whiskey jug. I did not like the whiskey – it was too strong after cream and strippings – but the old lady had a ten-gallon keg of grape wine in the making in the woodshed. The bunghole was covered with a piece of wire netting to keep out mice, spiders, flies, etc. So I got my straw again and sampled the wine. Boys, I liked it about as well as the cream and strippings. I drank too much one afternoon when the old lady went to visit a neighbor. When she came home she found me lying on the stable floor asleep. I played sick. Old lady gave me a big dose of salts. Not so good. Sent me to bed without any supper, to give the salts a right of way. After that I just got on and off the water wagon about every other day; but before I left for home I put a three-gallon pail of water in the wine keg, also two quarts of the old lady's whiskey, with the idea of balancing my wine budget. It must have balanced, for I heard afterwards that the old lady said that was the best keg of wine she ever brewed. As time and tide wait for no man, or little boys, and I had overstayed my leave, my brother David was sent to bring me home to go to school. I was loath to go home, as up to that time it was the only place I had boarded where I put on flesh. I went home from the old lady's like a fatted calf, with a full belly, a pair of new shoes, two skeins of yarn for my mother to knit me socks and mitts for winter, enough wincey to make a shirt, and fifteen cents in coppers as a bonus for being a good boy, and told me not to spend them foolishly but to drop them into the collection box each Sunday until all were gone. Oh, yes, also enough print to make mother a dress – a present from the old lady.

A lot of water has passed under the Saltford bridge since those days. My mind often wanders back to the dear old lady, the old clay pipe of peace, the cows, the cream, the strippings, the wine, the witch that stole the cream and strippings.

About this time the old saw-mill at Sheppardton changed hands. John Morrish bought the mill. Father remained as head sawyer. Also about this time D. E. McConnell came to teach the Sheppardton school. He taught this school for three years – '77, '78, '79. I got promoted into the Third Book under this teacher. I seemed to learn more from his teaching than from all the rest of the teachers put together. He did very little whipping. That was quite a change. He seemed to have the knack of making you learn whether you wanted to or not. He was a jolly good fellow and played ball, shinny and bull-in-the-ring with the boys. I guess that is why we tried to learn and please him.

I remember an amusing incident that happened at Sheppardton school. The teacher was giving the second class a lesson in natural history. There was a boy in the class who was very polite, like the boy in the other lesson of the Ox and the Ass and the Hay. Teacher says, "Well, now, Frank, can you tell me what kind of bird it is has a red head, speckled feathers and gets his dinner out of old stub trees, and you can hear hammer, hammer with his beak on the old stub tree?" Frank says, "Yes, master; that is a timber doodle." He thought it impolite to say "wood-pecker," so substituted "timber" for wood and "doodle" for pecker. It brought down the house, so to speak – made fun for the scholars, and the teacher laughed. But poor Frank – "timber doodle" stuck to him for a nickname for years. If any of the old scholars of those days see this in print, they will recall Frank, as he was a jolly nice boy, always spick and span and very polite. He lives in Toronto at present, is a grandfather, and the world has been very good to him. When he comes to Goderich he always visits me. I have a very warm spot in my heart for old "timber doodle" of the old school days, as he was the best man at my wedding. I trust Frank finishes the rest of his earthly journey in peace and comfort.

The last teacher I went to was Miss Yates – later Mrs. Harry Hayden. She did not spare the rod and spoil the child. I was much bigger than she, but she gave me several good whippings with a blue beech gad. I should be ashamed to tell it, but it is true. I often mentioned it to her in after years. She would laugh and say I richly deserved all I got. I guess she was right. After being promoted into the Third Book I went to school for a short time in the winters only. So at the age of eighteen I graduated with Third Book honors, if such there be. The Vision of Mirza was my hard lesson for three winters. The difficult words in it I could never learn, and where Joseph Addison found them all was a mystery to me: such as "inexpressibly," "apprehensions," "consummation," "contemplation," etc. I had to hunt up the old Third Book to learn to spell them here. But the old Third Book had some nice pieces, such as "Lucy Gray," "Speak Gently," "John Gilpin," "Solitude of Alexander Selkirk." There was also "The Old Arm Chair:"

> "I love it, I love it, and who shall dare
> To chide me for loving that old arm chair?
> I have treasured it long as a sainted prize,
> Have bedewed it with tears and embalmed it with sighs;
> 'Tis bound by a thousand ties to my heart;
> Not a tie will break, not a link will start.
> Would you learn the spell? A mother sat there.
> A sacred thing is that old arm chair."

"Bingen on the Rhine," too, I remember, as well as "Twenty Years Ago," beginning:

> "I wandered in the village, Tom;
>   I sat beneath the tree
> Upon the school-house playing ground
>   That sheltered you and me."

What boy or girl of the old Sheppardton school, or almost anywhere along the eastern shore of Lake Huron, will forget the 5th of September, 1881, the Dark Day? It started to get dark at noon, and at 2 p.m. the hens went to roost and people had to light the lamps. The school children went home, as there were no lamps in the schools. Everything seemed to take on a yellowish green tinge. The heat and air were stifling. Cattle, horses and sheep made for the bush or stable. They seemed to know something uncanny was going on. Many people, old and young, thought the world was coming to an end. I was very much scared myself and began to wish I had stayed with the preacher business, as I thought they were the only people who could go to heaven if this was the end of the world. As ashes seemed to be falling, my father said, "Michigan must be on fire," since the wind came west from the lake. So father and Billy Morrish took a lantern and went down to the boundary to the lake, about a mile and a half distant. When they came back they were certain the darkness was caused by fires in Michigan, as the water of the shore of the lake was covered with ashes. Well, this report eased the minds of most of us; but I was scared all night and until the sun got up next morning. I said my prayers that night, if I had neglected saying them every previous night. It rained in the night, with lightning and thunder. If you had seen the change next morning! Everything a dirty yellow. Creek was running with water that looked like the beer made at Wells' brewery at Slabtown. All the fish in the old boundary creek were killed. A little girl, named Fanny Burrows, when it began to get dark was sent for the cows at the back of the farm. She started the cows homeward, but they got scared and made for the bush. Fanny followed them and got lost in the darkness. Not returning at the expected time, search was made for her by her father and brother, who found her in the bush. Have heard it said that some of the natives (as well as myself) said their prayers on the Dark Day that had forgotten to say them for a very, very long time, so the Dark Day on Huron shores may not have been all in vain.

Now, when I left the old school and the old Third Book upon the shelf at the age of eighteen, I went forth into the world with a strong body but with very little education as taught in books; yet as I have travelled the road of life for threescore years and ten I have found the old world has been very good to me, and I only

HATTIE ANDREWS AND HER CLASS OF 1892 are shown in Sunday best at the old Sheppardton school where Gavin spent the happiest of his school years. As he passed by this site in the 1930s, he seemed "to see the boys and girls playing the same old games of leap frog, shinny, pump-pump-pull-away, bull in the ring; the teacher ringing the same old bell calling us in." He lamented, "But alas, the teacher sleeps upon the hill, and the pupils, some are in the churchyard laid, some sleep beneath the sea, and few are left of all our class excepting you and me."

GRADUATION FROM SHEPPARDTON SCHOOL was finally achieved "with
Third Book honours" by Gavin at age 18 in the school year 1879-80.
Shown reflecting the stern countenance of their teacher is Miss Long's class at
the Sheppardton school in 1915.

wish I may have as easy travelling in the next. Perhaps, dear reader, your thoughts may be carried back to your own youthful days when your mother dressed you up in your best, kissed your cheek and sent you off to school on the first day. With pride she watched you walk away; with anxious eyes she looked for your return. Oh, the sweetness as we think of the happy hours we spent in childhood and school days, of home and mother! Then the sadness to know they can never return, since we pass over the road only once. The first day at school – it is the first milepost on the journey through life, let it be long or short.

In closing let me say I have tried to give a correct account of my childhood and school days as I remember them. All the names I have used I have put down with due respect to those that are still alive, and in all honor to the memory of those that have finished their journey upon this earth. Now a fond farewell to all my old school teachers and old school mates until we meet again, I hope, in the sweet by and by.

YE OLDE CURIOSITY SHOPPE was the site where, in quiet moments, Gavin Green pencilled anecdotes and recollections on the backs of old calendar pads, letters, handbills and legal papers. Gathered together, these collections became the manuscript for his books. He is shown here at the door of his second shop "at the age of three score and ten, and then some."

Sketches Of Pioneer Days

Notice to Dust Kickers Who Visit This Old Curiosity Shop

WHY, yes, I know that everything in this old shop is covered with dust. Were not your ancestors manufactured from dust? You have no proof to offer that will hold water that the dust you are kicking about is not the dust of your ancestors, a King, Queen, a Caesar, Cleopatra, or a Huron Old Boy. For are we not told upon good authority that unto dust thou must return? Scientific writers assert that since the beginning of time the number of persons born upon the earth is 9,768,479,864,798,437,697,481,798,432,769,456. Divide this by the number of square miles of land upon the earth and you will see 1,927 persons lie buried in one square rod. So you will see this old earth has been dug over 128 times to bury its dead. So you may see by putting on your specs and figuring our dust question by the rule of three that the dust of your own sweet self some day in the future may decorate some ladies' chamber furniture, or some old tea kettle in some old junk shop, or be beaten out of some old rag carpet to drift around upon the highway of time, until you are overtaken by the water wagon of heaven and turned into mud, then back to the land you go again as dirt to grow turnips, carrots, potatoes and spinach to help feed posterity.

Lastly, but not leastly, you will be turned into gold dust, then into dollars to pay the butcher and the baker, the fiddler and the preacher, the doctor and the lawyer; and if you escape from the lawyer and the gold standard, you will be used to pay the war debt.

This is very sincerely for you,

G. H. GREEN,
April, 1932.
Goderich, Ontario.

HAMILTON STREET
GODERICH

1902

OPENING THE DOORS of Ye Olde Curiosity Shoppe for the first time in 1902
was an exciting event for Green and the townsfolk who soon gathered on the
storefront boardwalk to visit and share tales. The shop is shown here
in its first year of operation.
With Aggie in the doorway and Gavin standing proudly on the far right side,
they visited with the young barefoot boy, Fred and his father John Robertson;
Albert Paltridge; Dugald Morris; Harold Warrener, by the post; and,
Michael Ohler, sitting next to Green.

The Old Sheppardton Church

FIFTY years ago, in the gay eighties, this town of Sheppardton could boast of four business places: R. T. Haynes, general store and post office; William Bennett, general store; Frank Hathaway, blacksmith shop; and the Royal Oak hotel, kept by George Hilton. Two private dwelling houses, Randal Graham's and Richard Bennett's, were in the town limits. The suburbs — why, yes, the old town had its suburbs. To the west was the Orange Hall, with a large membership in those days, where the Gunpowder Plot was celebrated every 5th of November with an old-time dance, Billy McPhee and George Armstrong being the fiddlers. To the east of the town, up the boundary, were the school house, George Burrows' farm, and Morrish's saw-mill, also a blacksmith shop and a general store belonging to the Morrish mill, this eastern suburb in those days having a larger population than the town proper. Some of the families of those days were the Campbells, Bennetts, Burrows', Hawkins', Postelethwaites, Morrishs, Fishers, Vanstones, Greens, Simpsons and Fosters. John Walters, late of Saltford, was book-keeper and clerk in the store. Bobby McDonald, of Dunlop, was lumber and log scaler. Paul Morrish was the blacksmith. This was the extent of the town and its suburbs in 1880, when the last of the old landmarks was built — the Methodist church.

The Rev. Luther Rice, preacher of the Nile circuit, was the "father" of the church building. Along with him were Thomas Graham, James Graham, John Echlin, John Morrish, Frank Hathaway, and many others. Charles Hawkins was contractor for the building, George and Thomas Christilaw were carpenters, William Garside, of Goderich, was the painter, and the plastering, I believe, was done by John Sproul. The Rev. Luther Rice was the first preacher; John Echlin and Frank Hathaway were the first superintendent and teacher of the Sunday school. Also there were two young circuit riders who officiated at times. One of these was the Rev. J. T. Legear. I remember he travelled on horseback. Afterwards he married a daughter of David Fisher, and later moved to the State of Michigan. The other young preacher, the Rev. Mr. Foxen, also moved to Michigan. In addition, many local preachers in the early eighties used to expound the Scriptures to the natives. There were Charles Girvin, Joseph Hetherington, William Pellow, John Washington and others.

I believe all those so far mentioned have passed to their reward in that land that is fairer than day, with the exception of Robert McDonald, of Dunlop. John Morrish, the one-time mill-owner, died at the home of his son in the State of Dakota at the age of ninety.

Now to the passing of the old church, the last landmark of those bygone days, which was sold by auction on May 16th, 1932. For over fifty years it had served as a place of worship for the natives, and many of their children who went out into the world to do battle for a livelihood got their religious training within its walls.

As I journeyed out from Goderich to the auction sale, to see the last of the old landmarks of my old home town pass into other hands, a feeling of sadness crept over me. The clouds were heavy, with rain threatening, and quite a crowd of the natives and others had gathered to see the old church go under the auctioneer's hammer. We all went inside the church. The people

THE OLD SHEPPARDTON METHODIST CHURCH (C. 1880s) "where both saints and sinners went to worship: was sold by auction in 1932. Many of its contents dated back to 1856 and were removed from the Methodist New Connexion (sic) Church of Goderich when it was closed in 1878 to amalgamate with North Street United.

seemed to feel they were in a church, as in days gone by, as there was a sense of reverence, as well as sadness, in the air, and Auctioneer Gundry, with all his tact, could not raise so much as a smile upon the faces of the gathering. The sale started. The two pulpit lamps were knocked down to yours truly for one dollar. As they had served their day in lighting the pulpit for the preachers at their evening services, I left one with one of the old Sunday school teachers; the other I will keep myself as a remembrance of the old church.

The old pulpit next. I bid fifty cents. Albert Goldthorpe, Reeve of Colborne, raised the bid to one dollar. So the old pulpit went to the Reeve of Colborne. I had intended to buy it, but I thought if there was any virtue in the old pulpit they needed it in Colborne council chamber worse than I needed it in Goderich as a relic; so I let the Reeve of Colborne have it at his bid of one dollar. Who knows but it may still carry with it a spirit that may influence the Council and the taxpayers of Colborne to pay due respect and honor to the Reeve as he holds forth from behind it?

Who knows but the spirit of some of the old divines may inspire the Reeve of Colborne to expound the Scriptures to the Council and taxpayers, with such texts as "Blessed are the peacemakers," or "Render unto Caesar that which is Caesar's." For the old pulpit came originally to the old church from the Methodist New Connexion church of Goderich, built in 1856, which is now part of the old Doty Engine Works building. So we may see what influence the old pulpit may have in the council chamber of Colborne township.

The old pulpit railing, or communion railing, was knocked down to James Reynolds of the county jail for one dollar. There must have been some influence in the air to prompt a good Catholic like Jailer Reynolds to buy the communion railing out of a Methodist church. Perhaps it still has some virtue and may still do some good in the county jail. It also is a relic of the old Methodist new Connexion church, and at its railing many a sinner sought forgiveness of his sins.

The old organ that used to peal out its sacred melodies was sold for $6 to Thomas Dougherty, one of the original natives. It has gone to a good Anglican home after cheering the hearts of old and young with sacred music for many years. The pews (except some sold to Jailer Reynolds) went to Mr. Fry, of Detroit, who has a summer cottage on the bank of Lake Huron at Menesetung Park. Many a handsome girl, many aged fathers and mothers, as well as some bad boys, have sat in those pews since 1856, as the pews, in addition to the pulpit and the communion railing, came from the Methodist New Connexion church. May they be put to good use in their declining years, and may much joy and comfort come to those who occupy them in days to come. They are a very poor set of pews to sleep in – too straight in the back. I know, for I tried them out when I was a boy.

The old church building was knocked down to Mr. Fry for $150. I believe he will use the lumber to build summer cottages on the lake bank for his children and grandchildren. He will find the lumber first-class, for as boys my brother David and I helped our father to saw the lumber at the Morrish mill. We were only boys then, my brother and I, but we had to work for a living in our boyhood days. So you see I have a kind of reverence for the hemlock lumber in the old church, and may the use Mr. Fry puts it to bring him peace and prosperity, and may he enjoy his houses for many years.

The old church had many scenes, some sad, some joyous, within its walls in the old days. When the death angel came for a member of the congregation the pastor always preached a funeral sermon. This was always a very sad occasion to us boys and girls in our teens. Then there would be the revival meetings, when many of the old patriarchs from miles around would endeavour to convince the sinners to seek repentance of their sins. There were some fine men among these exhorters, such as Joseph Hetherington, John Echlin, Thomas Stuart, and others. They had some quaint and homely sayings, but their discourses were honest and sincere. I honor their memories more as I grow older, and many a boy and girl went out into the world from those revival meetings with a new vision of heaven.

Then the old-fashioned tea-meeting. I remember the one at the church opening. The boys starved all day so they could get the worth of their twenty-five cents in pie and cake, which were luxuries in those days. The motto was, Eat all you can, then fill your pockets to eat on the way home or take to school next day. Then the next night there was a ten-cent social to eat the fragments – which was the best of all for us boys. At this tea-meeting and church opening I was one of the boys chosen by the committee to act as one of the waiters. I was an overgrown country gawk, and, having been brought up on the Shorter Catechism and oatmeal, or, in other words, being of the Presbyterian persuasion, I regarded it as a great honor to be invited to be a waiter at the opening of the new Methodist church. Well, I strutted up the boundary to the suburbs at

Morrish's mills feeling like a young rooster wearing his first tail feathers. I had quite a time finding a girl partner; I asked three before I was accepted. However, I got through my first public function as a waiter at a tea-meeting, and strutted home with my tail feathers still in the air. Miss Emily Johnston, of Goderich township, was the school teacher at Sheppardton. At the tea-meeting she recited "Curfew Shall Not Ring Tonight," and the applause she got made the lumber in the old church ring. She afterwards married John B. Graham, and they are the parents of Dr. Meredith Graham, of Goderich.

As I cast my eyes over the crowd at the sale I saw very few of the old boys who went barefooted to school at the time the old church was built. There were Thomas Dougherty, John B. Graham, Nelson Graham, William Hawkins, John Foster, William Johnston, Angus Gordon, my brother David and I. Chief Postelethwaite and Joseph Wilson of Goderich were others of those barefoot boys of the time of which I write, but they were not at the sale.

I have a very warm spot in my heart for the old home town, the old church, the old preachers, and the old local preachers, the old Sunday school teachers, and for the old inhabitants of the old town and its suburbs. For the natives who worshipped in the old church and who have passed on to the better land I have a reverence and sweet memories of them as I knew them in the long ago. To those who worshipped in the old church and who are still upon this earthly highway, and, like myself, nearing their journey's end, I would say, May the good influences that came into their lives from the old church and its preachers and teachers follow them to the end of their journey. If they see this scribble and it awakens any sweet memories in their souls, I shall feel that I have not scribbled in vain.

For myself, I was brought up a Presbyterian and when a boy went to St. Andrew's church at Port Albert. In the good old days of which I write the Anglicans, Presbyterians, and Methodists lived in harmony and the natives patronized all the churches. The boys generally went to the church where there were the most pretty girls. For instance, in the old Methodist church were Lena Graham (afterwards Mrs. Judge Johnston), Lily Morrish, daughter of John Morrish (now Mrs. Smith, of New York City), and Ada Haynes, who married Henry Echlin.

Now, if you travel northward along the Blue Water Highway and go through the old town of Sheppardton after the church is wrecked, the only one of the original buildings you will see is the old blacksmith shop. The building on the corner where the old hotel used to stand was moved there a few years ago from a farm, after the old hotel was burned down. The suburbs, with their population, also passed away years ago.

I forgot to mention that one of the fair maidens of the old town eloped and married a young farmer from the Nile on the night of the opening tea-meeting in the church. She was handsome – and just sweet sixteen. So you see the old town had its romances in those days.

The Old Sheppardton Church in New Setting

THE old Methodist church that once graced my old home town of Sheppardton seems to be still upon the map of the county of Huron. A pilgrim from New York City, a native of Goderich, having read in his old home town paper the account of the sale of the old church, and having a boyhood recollection of the old church, built in 1856, was anxious to see, before returning to New York, what Mr. Fry was making out of the material of the old church building. So he got a friend to drive him and his wife out to Mr. Fry's and on his return called at my Old Curiosity Shop and told me that Mr. Fry was doing wonders with the old material. He advised me to go out and see for myself. So I broke one Sabbath morning in June and went out to view the beauties of Menesetung Park and see Mr. Fry's cottages on the banks of Lake Huron.

As I neared the lake bank I beheld a building that was a cross between a chapel and a bungalow. The siding on the "new" building had graced the old church at Sheppardton – the same old siding remembered from days when we boys who thought we were too big and important to go into the Sunday school would hang around outside the church until the small boys and the "goody-goody" boys came out. We worked crossword puzzles upon the old siding, writing our name and beneath it our best girl's name, then crossing out the corresponding letters to see if it came out odd or even. If odd, she would not marry us; if even, she would. Some of the bolder bad boys wrote letters and drew pictures on the same old siding.

As I neared the building I noticed the old windows through which the bad boys used to make faces at the good little boys in Sunday school. These were performances of the long, long ago, but I seem to be a boy again as these scenes pass before my eyes. But alas! All is vanity and vexation of spirit. The preacher discovered that frosting the windows prevented the bad boys from looking in to see if their best girls were there without going in themselves, and the result was more coppers on the collection plate. It also stopped the good men and women, the boys and girls from looking out through the windows to see who was going up or down the boundary road while the sermon was being preached. This was the preacher's innings, and he scored. I see the same old frosting on the windows as was there in days of yore.

The three windows that were on the side of the old church Mr. Fry has placed together in cathedral style on the east side, which lets the morning sun shine into a large reception room or hall. Directly opposite, on the west side, the other three windows are placed together in similar style, giving light from the western skies, as the sun sets it will reflect its beautiful rays upon these windows and soothe to rest those who are fortunate enough to

see through these windows the famous sunsets on Lake Huron.
When the evening shades gather and the sun has sunk below the
waters of Lake Huron, a large stone fireplace will warm and light
the room. A Gothic window from the church porch is on the
north end of the building, where the Northern Lights may reflect
their colors. At the south end is the other Gothic window, at the
entrance door.

If you are a stranger visiting in Goderich and wish to see
what I have been describing, take the Blue Water Highway north
after crossing the Maitland River and passing through the village
of Saltford. At the bottom of the hill as you leave the village, cast
your eyes heavenward and you will see the old Dunlop tomb
where the famous Huron pioneers, Dr. and Capt. Dunlop, are
buried. As you reach the top of the hill turn to your left and you
will arrive at Menesetung Park, and a few hundred yards up the
lake you will find Mr. Fry's home cottage and the one built from
the old church, which as the years roll on may become a shrine to
which Huron old boys and girls may make pilgrimage.

And why not Goderich have a shrine, for the lame, the sick
and those weary in spirit, as well as Midland, for tradition tells us
that the early missionaries sent by the Church of Rome were
murdered by Indians on the banks of Lake Huron near Goderich.
Who knows but it may be where this chapel cottage, built from
the old church, now stands, or near the spot. All will help to make
it sacred and holy and I am positive that nothing but good
influence can come from this building which is the offspring of
this old Methodist church and the memories of the saintly old
divines who ministered to the pioneers within its walls, and the
missionaries who perished by the Indians on Huron's banks. You
pilgrims, if you have a little bit of faith, half as big as a golf ball, let
your religion be Catholic, Protestant, Presbyterian, or nothing, if
you visit this building you will be enriched both in body and
mind, by the spirit of reverence that seems to linger about it.

Reminiscences Enjoyed in Far-Off South Africa

TO the Editor of The Goderich Star:

Sir: A copy of your paper has come to my hand containing an interesting article by your correspondent Gavin Green, telling the story of the sale of Sheppardton church. After over fifty years' absence from Canada it is particularly interesting to me, for the reason that fifty years ago I was the teacher of Sheppardton school and the incidents referred to and many of the names mentioned bring to my recollection three happy years in my early life. Gavin was a pupil of the school, and, incidentally, may I say the interesting way he tells his story flatters me in the belief that my efforts in teaching composition were not in vain.

"This letter may possibly come under the notice of some of my old pupils and may help them to recall memories, pleasant I hope, of the days when I tried to teach the young Sheppardton idea how to shoot. I was eighteen; it was my first experience in teaching; the school trustees were George Burrows, senior, Joseph Tigert, and Richard Haynes, all of whom I suppose have passed over to the great majority years ago. The school had about fifty pupils, some of whom were nearly my own age. The boys were easily manageable, but the older girls were at first difficult. There was a coterie of five who sat together; they were all nice girls, all good-looking and brimful of mischief. I was inexperienced and impressionable, and for a time I doubted whether I should be able to hold my job; but after a few weeks we began to know one another better and they settled down decorously and gently and peace, good-will and harmony prevailed. The intervening distance in time and ten thousand miles may protect me if I mention their names – Annie Burrows, Emma Graham, Ada Haynes, and Tilly and Lena Graham. They all bring back a shade of tender memories. Ada was a favorite with the young men of the neighborhood, and finally Harry Echlin, of the Nile, prevailed. Lena, the liveliest of the bunch, married a judge. I have lost trace of the others. Should my telling tales out of school evoke a personal letter of protest from any of these dear girls I shall be pleased to answer them.

"Of the many pleasant memories that come trooping back to me none is more outstanding than the unparalleled kindness and hospitality of the people, many of whose names were mentioned in Gavin's article. I constantly had invitations and visited a good deal, but retrospectively from this distance I can see that I had a predilection for visiting the homes where there were grown-up daughters. I remember on the first occasion of a visit I was asked by the eldest daughter to draw the outline of a cat on some material intended to be made into a door mat. I drew a fearsome-looking tom with its back bristling ready for a scrap. Then I was asked to outline the letters Y.M.C.A., one on each corner of the mat. I was perplexed. What could a Young Men's Christian Association do with a mat with such a repelling atrocity guarding its portals? Later, however, I was informed the letters stood for 'You May Come Again.' I came again, and often I came again!

"Among my older pupils were some who showed signs of histrionic ability worth developing, and in the winter we gave concerts with music, songs and recitations, ending up with a play such as 'Annabella's Poor Relations,' 'Pumpkin Ridge,' and others, the titles of which I have forgotten. These concerts always commanded a full house, and the star performers, including Johnny Graham, George Dougherty, David Green and the lively girl quintet before referred to, with some outside help, usually brought down the house.

"I held my job for three years on $300 a year, to me at that time a princely salary, and they were halcyon days now recalled with pleasure.

"Some of my old boys and girls who notice this letter may wish to know where I have been these fifty-odd years. For their information I may say my life has been a very eventful one and a very happy one. Two years after leaving Sheppardton I went out to Australia to organize and open up a branch of a Canadian publishing firm. After seven successful years there I went to England, and from London to the South American Republic for three years, as the representative of some English publishing houses. Whilst there I had the unique experience of seeing three revolutions. The only man I ever met showing a better revolution average was the editor of the Buenos Aires Standard, who had been in South America thirty-two years and had been in thirty-three revolutions. He had the advantage of me by a fraction. Deciding it was a good country to get out of I returned to London,

D. E. McCONNELL, "one of the distinguished old boys of Huron," was a teacher at the Sheppardton school in the 1880s. He writes fondly of memories during his early days as a teacher and shares his recollections about former students in a lengthy letter first published in The Goderich Star in 1932.

and after a business tour in Spain, where my knowledge of Spanish was a value, I was transferred to India and the Far East, including China and the Philippine Islands. After three years there I returned to England, spending six months in Egypt on the homeward journey, and incidentally visiting Palestine. After nearly twenty years' experience in many lands and on four continents I decided to make a permanent home in South Africa, where for twenty years I have enjoyed health, happiness and prosperity amid the beauties and amenities of the charming old city of Cape Town, where I have taken an active interest in its civic, religious and social life. In Sydney I married happily and my wife has been my ever-present companion in my travels, and she still gently pilots me along the paths of rectitude, and if spared for two more years we shall celebrate the fiftieth anniversary of our married life.

"My varied experience and my contact with many peoples causes me often to meditate and philosophize on this strange 'old curiosity shop' in which we live, but my conclusions are that it can be a delightful and happy old world to those who can adapt themselves to its vagaries.

"In the days when I first knew Gavin Green I was a young man who 'saw visions,' nearly all of which have been realized; now I am an old man beyond the threescore years and ten, and I 'dream dreams.' I thank your contributor for helping me to dream a pleasant dream.

(Signed) "D. E. McCONNELL,"

"Cape Town, S.A., July 15th, 1932."

A Visit to the Old Home Church Upon
Its 64th Anniversary

THE old St. Andrew's Presbyterian church, Port Albert, which passed into the United Church a few years ago, was where I got most of my early religious training, and having a very warm spot in my heart for the old church, whose anniversary services were being held on the evening of September 18th, I with my wife and two lady friends motored out to the church. One of the ladies, like myself, got her early religious training in old St. Andrew's. This lady, I am pleased to say, made better use of it than did your humble servant. Whether it was fate, or some unseen power, that directed her I know not; but she went as a missionary in Western Canada for a number of years and is at this present time one of the teachers of the young in Knox Presbyterian church Sunday school, Goderich.

Why did I go out? Well, to tell the truth, I guess I was like the little girl in Goderich who asked when Knox church Sunday school was going to reopen. Asked if she liked to go to Sunday school, she replied, "Yes, I want to see how the new basement looks." I wanted to see how the new basement in old St. Andrew's looked, as well as to see any other changes in the old church. Well, I could hardly recognize the poor old church of my childhood days. Where once the old church sat upon the ground it

ST. ANDREW'S PRESBYTERIAN CHURCH (C. 1868) at Port Albert is marked
in 1992 by a plaque and a cairn alongside Highway 21. The picture shows a
reunion gathering in 1932, before the church "wandered from the `faith of our
fathers' and joined the United Church."

is now perched upon the new basement, and thus is six feet
nearer heaven since she joined the union.

As I mounted the stone steps to go into the church, I first
entered a vestibule – yes, a real vestibule – the old porch that once
graced the entrance was gone. As I entered further, all I could
recognize of the old church inside was the pews, the same old
straight-back hard seats. They had painted them over and tried to
make them quarter-cut oak; but I knew the same old seats in their
new dress. But the old pulpit and the preacher's seat were gone; a
brand-new pulpit and pulpit chair adorned the platform – donated
to the church by Arthur Bennett, of Chicago, in memory of his

father and mother. Arthur, like myself, got his early religious training in this same old church. He and I were in the same Sunday school class. His father, John Bennett, was the superintendent, and James Quaid was the teacher. I remember more of what I learned in this Sunday school than of what I got from the preacher's sermons.

The windows – oh, what a change! I missed the old windows with their 10 by 12 panes frosted to keep the boys from looking outside, and where we used to write our names and draw pictures – all are gone, both the window frosting and the names, and in their place is pebbled stained glass, cathedral style, which gives the old church a kind of dignified and city appearance. I cast my eyes down and saw that the old church even has polished hardwood floors.

Now, if only we mortals after we reach three-score and ten could be made over like the old church, with a new basement, new vestibule, new windows, new pulpit, seats and floors, we could put on a dignified appearance and city airs as well as the old church. But all is vanity. A wise Creator has ordained it otherwise.

Oh, yes! I almost forgot the choir. In the old days the choir singers as they came in marched boldly up to the front and took their places on the platform, and gave the congregation lots of time and a good opportunity to look them over, criticize their dress and pass remarks upon their appearance before the preacher arrived. I remember four of the outstanding choir singers of the old days: George Colwell, precentor; his sister, Marion Colwell, afterwards Mrs. John Willis; Mrs. John Quaid, and her niece, Miss Mary Jane Smiley. In those days the choir sang nothing but psalms and paraphrases. Mrs. Quaid and her niece, Miss Smiley, must have belonged to the old Scotch kirk, for they would sing only the psalms of David. When the preacher, after the sermon, would give out a paraphrase as the closing praise selection, these two ladies would rise and leave the choir and march out of the church. But they were always on hand the next Sabbath. It seemed quite natural to the congregation for them not to sing the paraphrases. Well, I looked for the choir as in the days of my youth, and after a long wait I saw a man coming up out of the cellar carrying a lamp, which he placed on the organ. He was followed by a lady organist, who played a few nice selections; then up from the basement came the choir – a large number of very handsome ladies and younger girls, who filled the chairs, and all stood at attention and gave the congregation ample time to admire their beautiful form and dress. The ladies certainly did honor to the old church with their splendid singing, including an anthem. The men made their appearance in the same manner, but alas! Most of them, like myself and the old church, need to be made over again. However, the men all sang well. A lady in the choir, from Auburn, I believe, sang the Scottish song, "My Ain Country," so sweetly as to charm any old-time Presbyterians like myself in the congregation.

As I cast my eyes over the congregation I saw very few of the old Sunday school pupils of fifty years ago. I noticed Mary Dunbar, now Mrs. Angus Gordon; Bessie Bennett, now Mrs. James Hayden; Bessie Simpson, now Mrs. William Johnston; and my brother David, William Pellow and Joseph Wilson, like myself, visitors from Goderich. I noticed only three people who were young men or women in those days: Mrs. William McMillan, who was then Mary McKenzie; Mr. James Crawford and Mr. James Stevenson. There may have been others, but I did not notice them, as the light was poor and my eyes are growing dim. I must say the pulpit and platform were beautifully decorated with flowers.

Now the preacher. He was the Rev. William P. Lane, of the United church, Seaforth; and I went out to hear him preach and size him up and see how he stood up against his brother, the Rev. David Lane, of Knox church, Goderich. I also have a brother David, whom I have had to stand up against from my school days to this present hour. Of course, he is not a preacher. Now, in our family my brother David was always held up as a shining light and an example for me to follow, although I am his elder brother – as I believe Mr. Lane is likewise an elder brother of his David; if so, we are kindred spirits. I will just mention five – the old original David of the Bible, who killed Goliath; David Lloyd George, of World War fame; David McConnell, one of my old school teachers, now of Cape Town, South Africa; Rev. David Lane, of Knox church, Goderich; and, of course, my brother David. However, Mr. Lane preached an interesting, helpful, inspiring sermon, very appropriate for an anniversary. "Why go to church?" Do not throw the church overboard because you do not like the preacher, or because you do not like some of the congregation. This is a poor excuse. You should support the church for the institutions and things that the church supports. I agree with this doctrine and will try to live up to it.

I remember two ministers I used to like hear preach in old St.

Andrew's. One was a Mr. Cameron, who had an iron foot. I imagine I still hear the clatter of his metal foot as he walked up the aisle to the pulpit. He had a very handsome wife and she sometimes came and helped in the singing. The other was an Anglican, the Rev. James Carrie. There was a kind of charm to his voice, and he had a charming personality that used to keep us boys after Sunday school to hear him preach. The Anglicans used old St. Andrew's church for their services for many years before they built their church at Port Albert. Both these preachers lived at Dungannon.

Those two outstanding elders in the old church in those days were John Bennett and James Quaid. They were two of the finest men I ever knew. I imagine I see them yet taking up the collection in the old church. Each had a pole about eight feet long with a box on the end, which they passed along each seat for you to drop your coppers in. If we boys tried not to see the box as it passed we generally got a jab with it to remind us to drop in the copper, as we boys sometimes wished to save our coppers to buy candy. Yes, I myself was one of those boys.

And now to the memory of the saintly old men and women whom I looked up to when a boy, the early pioneers and staunch supporters of the old church as I recall them sitting in their accustomed pews every Sunday if at all possible. James Young, John McMillan, Mr. & Mrs. James Crawford, Mr. & Mrs. Andrew Quaid, Mr. & Mrs. Thomas Wilson, Mr. & Mrs. Arthur Bennett, Mr. & Mrs. Colwell, Mr. & Mrs. Donald McKenzie, Mr. & Mrs. James Mahaffy, Mr. & Mrs. Joseph Dunbar, and many others that I cannot recall, but not forgetting old Mrs. George Burrows and my own mother, Mrs. Peter Green, who both lived at Sheppardton and for fifty years walked from Sheppardton to the old church at Port Albert. If they were not sick or the roads impassable, they were always in their pews. They were both Presbyterians of the old school. These pioneers of old St. Andrew's church have all gone on to the home that is fairer than day. May their memories ever be kept green by those who try to follow in their footsteps.

Now, I have tried to write this with due respect and honor to all things sacred pertaining to the old church, and if I have erred I ask your forgiveness; and I trust what I have written may awake some sweet recollections of your childhood days, and that if you are nearing the threescore and ten you will find it balm of Gilead to your soul. I remain one of the old boys with tender recollections of old St. Andrew's church, Port Albert.

If you are interested in this old church and you happen to travel north from Goderich upon the Blue Water Highway, you will see this old church, one mile south of Port Albert, sitting all alone by the side of a creek.

October, 1932.

Strong Men of Colborne Township

THERE died in Seattle, Washington, at the home of his son Archie, in the year 1933, one Kenneth Morris, at the ripe age of ninety-four years. His ashes were brought back by his family and deposited beside the remains of his wife, Mary Rhynas, in the Colborne cemetery. This Kenneth Morris being about the last of the old pioneers, it awakened in my memory many of the exploits of the early pioneers as I knew them and what was told me of them by my father and others.

I will try and set them down to convey their original meaning, without showing any disrespect to those pioneers that have gone to that other land, or to their offspring that are still on top of the ground.

Now, Ken Morris, as he was called, was a big powerful man who stood six feet four inches in his stockinged feet. He followed farming, coopering, saw-milling, and various pursuits. He was a jolly fellow and made much fun for the boys. When he was a chunk of a boy he was working for his uncle, William Green, whose wife, Charlotte Morris, was Kenny's maternal aunt. His uncle had a well-digger digging a well. It was blue clay. Uncle Willie used to put the tape line down to the well-digger every night to measure how much he had dug during the day. Uncle Willie did not seem to be satisfied with the amount he dug in a day in this hard blue clay. So he says to the well-digger, "Dash your buttons! Kenny there can beat you digging; so dash your buttons! You run the windlass and Kenny will go down in the well tomorrow and dig." So that night, when Uncle Willie put the tape line down in the well to measure Kenny's work, behold you, Kenny had dug two feet more a day than the original well-digger. "Dash your buttons! I told you that Kenny could beat you at well digging." So Kenny was kept at the job until he got water at seventy-two feet. When pump and pump logs arrived from Henry Dodd's pump factory, according to Kenny's measurement there were fourteen feet more of pump logs than were needed. Kenny, to keep up his reputation as a well-digger, gave Uncle Willie extra measurement every night by pulling the tape line down one or two feet until the extra fourteen feet were measured. I believe Kenny was just a big barefoot boy of fifteen when he soldiered his uncle out of fourteen feet extra in his well digging (dash his buttons!). Uncle Willie was a good Green. He never put more hellfire and brimstone into his cuss words than he could get into "Dash my buttons." He always said family prayers night and morning, rain or shine.

Kenny's half-brother, James Cassidy, and my father, Peter Green, ran one of the first threshing machines in the township of Colborne. The old open cylinder type; in fact, the machine just consisted of a cylinder. One of the threshers fed the loose sheaves through the cylinder, one shook the straw when it came through to free the grain, the other man kept the horse going on the treadmill that furnished the power. This was the first threshing improvement on the old flail.

Another of Kenny's harmless tricks. One morning, early, he was going to Guy Hamilton's to work. He met an old gray mare on the road, so he jumped on her back and rode her down to Cracky's Corners (now Loyal), where Cracky Robinson kept the Plough Boys' inn. He opened the bar-room door and put the old mare in the bar. When Cracky came down to see what the racket

ANDREW GREEN, grandfather of Gavin, was "all bone and muscle" at 225 pounds. According to Gavin, he "carried a barrel of whiskey from the dock in Goderich to the Crown and Anchor Hotel at Gairbraid, one and one-half miles over a bush road, through the John Galt property." That's said to be 500 pounds!

was, here was the old gray mare waiting for a drink, with Kenny sitting on her back.

I have heard my father telling of Kenny eating a gallon of batter made into pancakes at one sitting. He came home to his father's one night; his stepmother was just going to make pancakes for supper. Kenny said, "Dall it, mam, I am very hungry." She said, "Sit in, Ken, and I will make you some pancakes." Aunt Lillie, as she was called, kept making pancakes, and Kenny kept eating them until he had eat eaten up the whole gallon of batter made into pancakes. Ken says, "Dall it, mam,

I was hungry. Nothing to eat all day."

There were many of the early pioneers who were big powerful men. I will tell you of some of their exploits. Some of these things I knew of myself; others I got from my father and neighbors. They are all true and founded on facts.

I will start with my grandfather, Andrew Green, who stood six feet high and weighed 225 pounds, all bone and muscle. He came from Perthshire, Scotland, in the year 1833, and was the father of the first white child born in the township of Colborne, Charlotte Green (later Mrs. James Anderson, of Brantford, Ontario). He also did the first ploughing in the township of Colborne, on the Maitland flats, for Dr. Dunlop, in 1834; was also

JOHN MORRIS, another of "the big, powerful men" of Colborne Township, "carried 100 pounds of flour, a No. 4 plough and some groceries" to the Morris settlement on the 10th Concession, "about five and a half miles."

the first constable appointed in Colborne township, and was at the first council meeting. He at one time carried a barrel of whiskey from the harbor at Goderich to the tavern at Gairbraid, which was situated at the crossroads, on the south-west corner of the farm now owned by James McManus. Carried it through the bush on a cowpath through the Galt property (now owned by Mr. Fleming), a distance of about one and a half miles. A barrel of whiskey weighed over 500 pounds.

Then there was John Morris, not as big a man as my grandfather, who carried 100 pounds of flour, a No. 4 plough and some groceries on his back from Goderich to the Morris settlement on the 10th concession, about five and a half miles from Goderich.

Then John Buchanan, who carried eleven bushels of wheat across the barn floor and back again. Six hundred and sixty pounds! He could have carried four bushels more if they could have put it on his shoulders, so he said. John also came from Scotland, and, like my grandfather, lived upon oatmeal until he came of age. I heard my grandfather say that he got meat once a year only, at New Year, and very seldom white bread when he was a boy. John Buchanan's first job was driving a team of horses for Sheriff McDonald, drawing wood from his bush farm across the river to Goderich. One day, coming up the bridge hill, one of

JOHN BUCHANAN "who carried eleven bushels of wheat the length of a barn floor and back" claimed he could have carried fifteen "if they could have piled it on him."

the horses got balky. John grabbed the horse by the bridle, hauled off and struck the horse on the side of the head with his fist. He must have struck in a vital spot, for the horse dropped, to rise no more. John had "killed him dead." Would not this John have played hob with John L. Sullivan, Corbett, Fitzsimmons and the old prize-fighters of other days, if he had got a chance to hit them? Now, the truth of all this is, John had to pay Sheriff McDonald for the horse. He had to work eight months at $12.00 per month to pay $96.00 for the horse. This same John had a son John, who was no slouch of a strong boy. He could take a barrel of salt and load it into a wagon by grabbing the barrel at each end and throwing it over the tail-board.

There was also big Anthony Allen, who kept the tavern at Dunlop. He stood six feet six inches in his stockinged feet and weighed 300 pounds. He was the biggest of all the old-time strong men. He could take two ordinary, everyday men by the coat collar, hold them out at arm's length, and crack their heads and heels together. Now, the wild, bad boys from Goderich town never came along to clean out Anthony's tavern. One of the pastimes of the bad boys from town was to swoop down on country taverns and take possession of the bar when they got liquored up. If you came drunk to big Anthony's tavern you would get no more liquor and out you would go – peaceably, too, or Anthony's big foot would follow you out.

ANTHONY ALLEN "who could take two men, one in each hand, hold them out at arm's length and crack their heads and heels together" kept good order in his tavern at Dunlop.

Another of the strong men was Big Bob Linfield. Big Bob was a native of New Brunswick. He sailed the Lakes with Captain Andrew Bogie on the Jeannie Rumball. One day at Captain Babb's hotel at the harbor, the different captains were letting off their hot air, about the exploits of the strong men of their different crews. Captain Bogie said, "I will buy the drinks for all the sailors between Hell and Kincardine, if my sailor, Big Bob, cannot lift and carry the anchor of the Jeannie across the deck." All hands left Captain Babb's hotel and boarded the Jeannie Rumball, expecting to return to Babb's and get free drinks. Capt. Bogie says, "Bob, lift the anchor." Big Bob grabbed the anchor, carried it across the deck and dragged the anchor chains along with it. I have been told on good authority the anchor weighed over 700 pounds. Sorry I have not a picture of Big Bob to insert here. He was six feet six inches in height and weighed 250 pounds.

The first Goderich horse races were held on the Colborne side of the river flats. As was customary in those days at gatherings such as this, there was always someone looking for a fight. Some of the bad boys from Goderich town across the river thought it about time to start a fight, so strutted around looking for trouble. Old man John Allen got down off the grandstand, which was a rail fence, and told those fighting men from Goderich town that his son Anthony could lick any damn man on the flats and that he could lick Anthony. So when these fighting men from Goderich town got a look at Anthony they began to think that old man John Allen was about right, and backed water across the river. I was told this by a man who was there at the time. Now, big Anthony was not a quarrelsome nor a fighting man. Quite the opposite. He always tried to keep peace between the neighbors. He acted as a kind of boss around Dunlop. Many of the neighbors went to Anthony for advice when in trouble. He was a shrewd business man, a good neighbor, and a good friend in need. But, like the rest of the old strong boys of Colborne, he has gone to the land where strength does not count.

Then there was George Currie, of Nile, who used to thresh with the old horse-power. George could lift one end of the horse-power high enough to allow the wagon to be run back under it to load it. I heard of only one other thresher who could do that. This was Jim Young, in Ashfield, near Sheppardton. Jim could lift one end of a ten-horse-power up for the wagon to be backed under. This was before horse-powers were mounted on trucks. Now, this George Currie had a brother Billy who used to draw sawlogs to Morrish's sawmill at Sheppardton. He could load the logs himself on to the sleigh without their being put on a skidway, simply by picking up one end, placing it on the front bob, then going to the other end and placing it on the hind bob –just like that. I remember one day Billy was coming to the mill with a load of logs. One of the traces broke, about two miles from the mill. No hay wire in those days. Billy was not to be beaten; he grabbed the broken trace and kept up his end of the whiffle-tree until he got to the mill with the load of logs. I know this to be true, for I was working at the sawmill at the time with my father. I also knew Billy to be a powerful man. He was bigger and stronger than his brother George, of horse-power fame.

Then there was James Sallows, a stout, strong, hardy man. He could chop wood in the bush all day, come home and do his chores, take his flail and thresh grain until 12 o'clock, and get up fresh as a daisy next morning. He, like the Morris', had a large family, eleven daughters and four sons. At present only one of the male descendants is living in the township of Colborne: viz., William Sallows, township clerk, on the old homestead. I have heard my father tell a story of how when the first pressed yeast cakes were sold in Goderich they were in squares put up in paper packages. Mrs. Sallows was in Goderich and bought a package of these yeast cakes and put them in a bureau drawer for safe keeping. Mr. Sallows was one day looking for something in this bureau drawer. He came across the yeast cakes, and thought they were some kind of special cake his wife had bought for herself when in town. He says to himself, "Dall, it will be a good joke to eat mam's cakes," so he ate the whole package, twelve cakes in all. He was such a strong man that the cakes did not hurt him, but he had to drink a gallon of water, so he said, to get the risings off his stomach after eating mam's dall yeast cakes.

Then there were, beside the strong men, the grain cradlers and wood choppers of those early pioneer days. There was David Bogie, a big powerful specimen of manhood. David felled the trees, chopped and split eight cords of beech and maple in one day in the month of April. This was chopped on the farm then owned by Thomas Morrow, situated on the boundary line, near Sheppardton. The farm is now owned by William Vrooman. John Dustow and William Morrow, both living today, are the men who helped pile the wood, and they can verify the truth of this.

Then there was George Bennett, of Sheppardton, son of Richard Bennett – Irish Bennett, as he was called. Another fine big specimen of a man. He could chop four or five cords of wood almost any day. George had a peculiar characteristic – he would not be bluffed. I remember one time at the little store at Morrish's mill, at Sheppardton, one of the boys, Harry Scales by name, saying, "I will buy you all the eggs you will suck raw." George says, "All right, bring on your eggs." John Walters, clerk, brought out a basket of eggs. George started, and he sucked three dozen

ROBERT BOGIE AND HIS WIFE are seated to the left of DAVID BOGIE, reputed to have "chopped and split eight cords of maple wood" and "cradled seven acres of grain in one day."

and one eggs raw without stopping. Harry paid for the eggs. George offered to suck another one and a half dozen if Harry would pay for them, but Harry declined. Another time, at the same store, one of the mill men said to George, "I will bet you a dollar you cannot take a whole box of Ayers' pills at one dose." "Done," says George. John Walters handed out the box of pills. George took the whole box at a dose, and a drink of cold water as a chaser to keep the pills in circulation. George survived the pills for many years, and died only a few years ago in Bay City, Michigan, where he went in 1880. I think he was foreman for a big lumber company there. I was at Morrish's mill at the time of the eggs and pills episodes in 1878, therefore I know this to be true.

There were lots of good wood choppers besides Dave Bogie and George Bennett, but this same Dave Bogie was a great grain cradler as well as wood chopper. He started out one day to try to cut eight acres of grain before sundown to put with the eight cords of wood he had chopped. However, David cradled six and a half acres of oats and called it a day. There was a heavy wind blowing and he could not carry his swath, so had to walk back, or he could have cradled the eight acres, he said. This cradling was done on the farm of Richard Morrow. John Dustow, William Morrow, and Richard Morrow bound the oats into sheaves. John Dustow and William Morrow are both alive today and can verify this to be true. This farm is now owned by Alex. Watson. Dave Bogie was the champion wood chopper and grain cradler of Colborne township. He, like George Bennett, left his native country and went to Glenwood, Minnesota, U.S.A., where he was marshal of the town until his death a few years ago.

Then there was Andrew Peacock, who lived with his uncle, John Donnelly, on the Ashfield side of the boundary line, near Sheppardton, who could cradle four to five acres of grain almost every day. Andy was a big raw-boned man, stood six feet three inches, and could cut a swath eleven feet wide with a grain cradle. He made his own cradle, called the turkey wing. Andy also emigrated to Uncle Sam's domains, and he, too, died a few years ago. All these old-time pioneers, the strong men, grain cradlers and wood choppers, have now gone to that land where, we are told, there is no lifting, cradling nor chopping.

Now, there were other strong men and wood choppers and grain cradlers in Colborne township besides those I have mentioned, as I only took in the lake shore part of Colborne, where I was born. Over in the Dutch settlement, the Devonshire settlement, and the Youngs' settlement there were many men of power and fame both at work and play. I might mention one Patterson, a great cradler up in the Youngs' settlement.

Now, in this part of Colborne, in the Bogie, Green, Sallows and Morris settlement, there was a word often used by the Morrises – especially Uncle Sam Morris – in their everyday conversation. It was either coined by them or came over as an emigrant from England. It was the word "dall," and it went like this: "By dall!" "Boys, get those dall horses and cows up from the dall pasture field before breakfast, or I'll give you all a dall good trimming." I often heard the word "dall," as I was often at Uncle Sam's, since he was married to my father's sister, Lillie Green. One day I asked my father what was the meaning of the word "dall" that Uncle Sam said so often. He replied, "Well, boy, it is just a polite way of saying 'damn,' but don't you say it." But I often used to say "by dall" when my father and mother were out of hearing distance, and I often say "dall" to this day, as it saves me from committing a worse sin by saying "damn." This word "dall" was often used by small children. For example, one of the Morris mothers sent her six-year-old son to mind his baby brother, aged two, while she made some smearcase (a kind of cheese) for dinner. Baby brother cried, kicked up a rumpus; the mother went out to see what was wrong. "What in the world will I do with you, Sammy, to stop your crying?" Little six-year-old says, "Dall it, mam! Slapen's dall backside, then he'll behaven's self."

The Morris family was one of the largest that emigrated to

Colborne township. They came from Westbury, Wilshire, England. Besides the father and mother there were nine children, six boys and three girls. The Morrises all followed the trade of weavers before leaving England for Canada. The boys and girls all married and raised large families in Colborne township. They were known through the Morris settlement as uncles and aunts by their own children and by their neighbors' children. There was Uncle John Morris, Uncle Prince Morris, Uncle Sam Morris, Uncle Alfred Morris, Uncle Mark Morris, Uncle Jim Morris. Of the three girls, Aunt Sarah married James Sallows, and Charlotte married William Green, and Hannah married Sandy Green. Their children and grandchildren are scattered over Canada and the United States and other foreign lands. There are only three men of the Morris name, offspring of the old original pioneers, in the old township of Colborne: Dougald Morris, a great-grandson of Uncle John; Abner Morris, a son of Uncle Sam, and Abner's son Kenneth. Abner and Kenneth live on the old Morris homestead. Uncle Sam Morris and his wife, Lillie Green, are the oldest married couple buried in Colborne cemetery. Uncle Sam died at the age of ninety-five and his wife, Lillie, at the age of ninety-six. Kenneth Morris, eldest son of Uncle Sam, died at the age of ninety-four, and Uncle Sam's sister, Charlotte, died at the age of ninety-six.

The old original pioneers, the Morrises, Sallowses, Greens, Bogies, Armands, McHardys, Buchanans, Clarks, Stewarts, Dustows, Morrows, McCanns, Thompsons, Youngs, and many others, have all passed to that land beyond, to which you and I are all people that on earth do dwell are journeying. To those who have gone and finished their journey, may their souls rest in peace and happiness. And to the offspring of the old pioneers of Colborne township who are still left on top of the ground and are journeying to that land where their forefathers have gone, may they, with myself included, all reach that land with a clean bill of lading.

December, 1934.

The Old Men

(by Ernest H. A. Home)

I often think of the old men
Sitting alone in the sun,
Watching the bees on the blossoms
Now that their work is done.
They are so wistful, the old men,
So wistful and gently wise,
My heart goes out to the old men
Whenever an old man dies.

I often sit with the old men,
So eager to speak are they
Of those by the world forgotten,
Of things of a bygone day.
They know so much, do the old men,
So much, though their books are few.
I like to sit with the old men;
The words that they speak are true.

I often walk with the old men,
Suiting my steps to theirs,

And I think, when our tongues are silent,
How many their griefs and cares!
But it's little the old men murmur;
They vision with clearer eyes,
And patient are they, the old men,
So patient and calmly wise.

I claim as my friends the old men
Sprung of our native soil,
Whose bodies are worn and feeble
And bowed by the years of toil;
And they, in their God-sent knowledge,
They read me with kindly eyes,
And see, when my words are halting,
The thought that behind them lies.

I keep a place for the old men,
A place for them in my heart,
For dauntlessly well and truly
Each of them played his part;
And now they are spent and weary,
For theirs was no royal way,
The old, old men whose labours
Are yielding their fruits today.

Oh, often I think of the old men
Sitting alone in the sun,
Watching the shadows grow longer,
Leaving us one by one.
They are so trustful, the old men,
So trustful and simply wise;
Heaven's richer and earth is poorer
Whenever an old man dies.

Christmas Holidays in the Sixties and Seventies

YES, the pioneers had a Santa Claus, but he travelled light. He was not overloaded with luxuries as he travels today. But joy, happiness and good-fellowship were in the air, and both Jack and his master ate at the same table in those good old days. You could buy a gallon of whiskey for one dollar, and a keg of Wells' beer for one dollar and a quarter, and if you wished to hog it and have a cheap drunk you could have one for about one York shilling (12 1/2 cents). All of these drinks you could have without a government permit.

The first Christmas I remember was '66. My parents lived near Dungannon, in the house where Bobby Hamilton lives at present. I had an aunt from Michigan visiting at our house this Christmas; she had a little girl named Sadie about my age. I did not know anything about Santa Claus then. Christmas morning I got some candy and nuts in my stocking and a toy called a jumping jack (you pulled a string and he sprawled his arms and legs around). Sadie got a nice little doll with a red dress and gold braid. Well, I wanted the doll. I wanted Sadie to trade and give me the dollar in exchange for my jumping jack. She would not, so I gave the jumping jack to my younger brother, David, and in about ten minutes it was all in pieces. I began a howling match, as they called it in those days, and to pacify me my father had to go to the village to Johnny Roberts' store and buy me a doll with a red dress. My mother and my aunt tried to shame me and called me a little girl, nursing my doll; but I stayed faithful to the doll with the red dress, and after threescore years if I happen to pass a toy shop window I stop and look for a doll similarly clothed. And if I see one of those dolls walking on the street – well! well! My fancy still runs to the doll with the red dress that I cried for on that first Christmas morning that I remember.

Before the next Christmas came around I found out somehow that there was a Santa Claus and that he came down the stovepipe, through the stove and out the stove door, to put presents in our stockings. Now, in those days we had one of the old high oven stoves called the Clinton Airtight. David and I could figure out that Santa could get down the stovepipe if he was small enough, but how he was to get around the old oven and out the stove door was a mystery to us. But David and I figured on Christmas Eve we would stay awake until father and mother went to bed, then we would get up and take the stovepipe off the stove, so that Santa would not have to go through the stove; but kind nature stepped in and we dropped off to sleep, to find to our surprise when we awoke in the morning that Santa had been there and filled our stockings. But how Santa got around the oven of that old high oven stove was a mystery to David and me for years.

I remember David and me getting from Santa Claus that Christmas a juvenile book each with the A, B, C's in colors and each letter represented by characters in colors, with a rhyme for each letter of the alphabet. We learned these off by heart, and I remember the rhymes to this day. It may amuse some of the younger children and may refresh some of the older people's minds and carry them back to their childhood Christmas days, so I will insert them below as I remember them:

A was an Archer that shot with a bow,
B was a Beggar with tale full of woe,

C was a Candyman that sold lots of sweets,
D was a Drunkard that slept on the streets,
E for an Elf that danced with a fairy,
F was a Fox both cunning and daring,
G was a German that drank lager beer,
H was a Hunter that just killed a deer,
I was an Indian that shot with his bow,
J was a Juggler that made a great show,
K was a Knight that carried a lance,
L was a Lady learning to dance,
M was a Music man that played a nice tune,
N was a Negro chasing a coon,
O was an Ostrich said to eat stones,
P was a Panther that gnawed up the bones,
Q was a Quaker that wore a broad brim,
R was Rebecca his wife that was thin,
S was a Stag, a species of deer,
T was a Turk born without fear,
U was a Union boy that carried a flag,
V was a Veteran that had but one leg,
W was Winter, the season of snow,
X was Xertis, who had nowhere to go,
Y was a Youth that smoked a cigar,
Z was a Zouave that had been to the war.

One Christmas, I remember, we lived at Dungannon, as my father worked for Thomas Disher in the woollen mill. Mr. Disher's ancestors were Dutch, and wishing to keep up the traditions of his Dutch ancestors, he had a suckling pig dressed for Christmas. I remember father taking David and me over to Disher's to see the table spread for the banquet. I remember the little pig standing in the middle of the table upon a long platter, looking as if it was alive. Now, don't mistake me; we were not at this banquet, but Scotch William McArthur and some of the other aristocrats from the village were there, also Mr. and Mrs. Richard West from across the river.

The week before Christmas generally was a kind of holiday week at the public schools. On the Friday before Christmas there was examination at the school; trustees and parents came to see what progress their children had made during the year. The trustees and preachers examined the pupils, and prizes were awarded as to standing, every pupil getting a prize according to his merits. The prizes always consisted of books, and they were highly prized in those days. Some of the smarter pupils gave recitations and dialogues, and trustees always gave the pupils a treat of candies, nuts and apples, which were also a luxury. This put the Christmas spirit in the air and everybody seemed to be happy.

It was customary in those days to have a jug of whiskey at Christmas to treat your friends when they called; my father generally had a jug. Also my father always had to have a treat of a can of oysters at Christmas holidays: they came in a tin can sealed like a varnish can. I remember that there were the words "Packers, Baltimore," on the cans.

My parents being Scotch, we generally had a celebration on New Year's as well as Christmas. Mother always made currant bread for New Year's, and when neighbors and friends called they always got a piece of currant bread with the customary glass of whiskey. Some of the richer classes furnished shortbread with the whiskey.

During the holiday week not much manual labor was done; the time was given over to visiting friends and neighbors, and sports, playing cards at night, sleigh riding parties for the young people, dances at private homes and at public dance halls – and nearly all country and village hotels had a ball room for dancing. I have attended many of these old-time dances, both at private homes and at public halls, and there generally was plenty of beer and whiskey, but I never in my time saw a young girl at any of these dance parties take a glass of whiskey or beer. It would have been something shocking for a young lady to indulge in intoxicating liquor, and very seldom did any of the boys get the worse of liquor so that they had to be thrown out. There was always someone capable of throwing you out if you did not. behave yourself.

Some of the old-time fiddlers of those days who played for these Christmas dances were Billie McPhee, George Armstrong, Leslie Currell, David Hale, Albert Thurlow. Billie Lasham, Jr., of Slabtown, was a young nifty, classy fiddler who generally played at high-toned balls and weddings. I heard him play at my Uncle William Green's when his daughter, Annie, was married to Wattie Watson, of Goderich, when I was a boy.

Then there were the customary shooting matches for turkeys and geese, rifle shooting and shotgun shooting. Shotgun shooting was generally from thirty to fifty yards at the white card, ten cents per shot, and usually ten for each fowl. Whoever put most shots in the card got the turkey or goose. The old-time shooting match that I speak of was at Sheppardton. Bill Lasham, Bob Ellis, Dick Fritzley and others from Slabtown came up and got their share of turkeys and geese; but they were good sports, as they generally brought along a keg of Wells' beer, all free for the boys. Captain Andrew Bogie, Captain Jimmy Green, and Jimmy McHardy, three sailor boys, and Joe Thompson generally got their share of the turkeys and geese at these shooting matches.

Rifle shooting was also at a white card, usually one hundred and two hundred yards distant, and the one that came nearest to the bull's eye got the goose and turkey. I remember Billy Young was one of the crack rifle shots. He shot a padlock off a barn door at 100 yards. One day he was going home from hunting near Port Albert. He spied a dog running across a field about 300 yards away, and Billy thought he would take a shot at him, not expecting to hit him. Well, he did; and that crack shot cost Billy the price of a cow. The owner of the dog sued Billy, but Billy settled; he said it was the best shot he ever made, if it did cost some siller.

Then there were the old-time hunting and shooting matches, where two parties chose captains, about twenty to each side, on a certain set day, starting at 9 o'clock in the morning and quitting at 4 p.m. The parties chosen walked to the bush and hunted inside of a certain limit. Partridges scored 30, rabbits 10, coons 40, foxes 100, black squirrels 10, red squirrels 5, chipmunks 5, woodpeckers 15, and so on, and whichever side had the game that scored the highest won. The losers had to pay for an oyster supper for the winners and for fiddlers for the ball at night. Sometimes this supper was held at Point Farm; this last one was held at Sheppardton hotel, if I remember rightly. Joe Thompson, of Colborne, was one of the captains; Thomas Johnston, of Ashfield, was the other captain.* This was one of the last shooting matches that were held in Ashfield or Colborne townships. It was cruel sport, for many an innocent little squirrel and bird that came out to view the Christmas scenes never again returned to its native haunts.

Now, as most of all these incidents happened threescore years ago, and most of those I have mentioned have gone to spend their Christmas in another land, there is a sweetness creeps over me as I think of the many happy Christmases I have spent upon this earth. Through the process of time there cannot be many more. A sadness also creeps over me when I think of those Christmas seasons I might have made more joyous for many a poor creature upon the road of life. So, dear reader, as you read this, and if you are still in your youth, do not let a Christmas season pass by without putting forth an effort to make some poor and lonely person happy upon a Christmas Day. Then, if you live to be threescore and ten, you will have nothing but sweet recollections of thought to carry you back over your Christmas seasons spent upon earth, and much joy and happiness will go with you until you depart to spend your Christmases in that land that is fairer than day.

I forgot to mention that in those old Christmas days they did not greet you as we do today with "Compliments of the season," or "Merry Christmas." It was this way: "Christmas box on you, Tom" (or Mary), and the one that got the first crack at the box won. The other was supposed to give a present of a box or some other article.

So Christmas box on you, one and all!

*P.S. – Both captains of this shooting match have gone to the happy hunting ground since this was penned.

In the Days of the Linen Duster, Bustles and Hoop-skirts

THE romance and sport writers of 1936 seem to draw their soapy lather from the gay nineties. How about the gay seventies and eighties, when the young men and maidens of those pioneer days began to wear store clothes and gaiter shoes? A boy so dressed was generally called a dude by those who could not afford store clothes and gaiter shoes and had to be contented with homespun clothing and the long-legged cowhide boots. The store clothes young man usually sported a long coat made of linen, which reached almost to the ground, called a linen duster. This was to save the store clothes from the dust and mud, flies, bugs and other hairy animals that infested the dirt roads in the seventies. This was about the first Paris fashion taken up by the young people of pioneer days. It also spread to the older pioneers who had a touch of the aristocrat in their blood.

The first linen duster I remember seeing was worn by a school teacher, R.T. Haynes. He later kept the store and postoffice at Sheppardton, my old home town. Haynes could walk from Sheppardton to Goderich, eight miles, in one hour. On a hot day, when he had steam up, you could see him coming up the Lake Shore road and the old linen duster flying out behind like a pair of wings that the eagle wears on the Yankee silver dollar. I also

AGGIE AND GAVIN set out in search of "a bright future on the matrimonial sea" according to the newspaper article which reported their wedding in December, 1892. They are shown here in "wedding toggery of the gay nineties."

remember big Joe Williamson, who stood six feet six inches, wearing one of these linen dusters. He was a kind of jolly good boy and a home-made poet, one of his poems being "Dungannon's Lovely Girls." Donald McNevin recited this poem to us one day in Sheardown's blacksmith shop. Donald said, "Boys, this is true, for I married one of the Dungannon Lovely Girls myself." (Since the above was penned my old school boy friend Donald has passed on to the world of realities, where I trust he may meet the friends of youth and Dungannon's Lovely Girls.)

My grandfather Green was one of the other old pioneers I remember wearing the linen duster. But I guess he wore it for dignity or fashion, as I never saw him dressed in store clothes. For myself, I could not sport store clothes and a linen duster, but I came in on the paper collars and hair oil. You could buy a box of paper collars, one dozen in a lot, for a York shilling (12 1/2 cents), and a bottle of very highly perfumed hair oil for ten cents. Before going to church, picnics, dances, or to see our best girl, we always put on a fresh paper collar and a liberal supply of hair oil, which went a long way to touch up the home-spun clothes and cowhide boots. The poor paper collar generally went the way of all flesh and grass, back to the land, after getting soaked with hair oil and sweat. Yes, we did not just perspire: we sweat in the good old pioneer country dances. The paper collars cost only one cent, and there was no washing and no ironing for our mothers to do.

I remember when I was a chunk of a boy in my teens I was invited to a country dance at a neighbor's. I had the paper collar but no hair oil, so I went to the goose grease bottle, but it was empty; my brother David had the croup and mother had used up all the goose grease on him. The next best lubricant to goose grease was pork gravy. Well, I got two tablespoonfuls of pork gravy, one spoonful of ground cinnamon and applied it to my hair. Didn't my hair shine! Yes, my hair looked and smelled like a billy goat coming in out of a rainstorm. To the dance I went, and as I warmed up dancing the gravy began to drip down my neck and soon spoiled my paper collar, which I had to discard. But I could not get rid of the pork gravy and cinnamon so easily. I brought some of it home with me, and the pillow-case on my bed got a good share of that gravy and cinnamon hair oil.

However, whether it was the hair oil, goose grease or pork gravy, or the little fellows that played tag in our hair in those pioneer days, I never remember seeing any bald heads or dandruff. Boys and girls all had heavy crops of hair on their heads; and, as I was a kind of pioneer amateur barber, I was often called upon by the neighbors to cut the children's hair. I might just say in passing that Chief of Police Dick Postelethwaite was one of the boys I practised shingling hair on. Well, Dick had a fine head of hair and a good tough scalp. He never squealed when the scissors pulled or nipped. Never saw any dandruff in Dick's hair in those days. Dick and I still have good crops of hair on our heads, minus the pork gravy and the little fellows that looked after the health of our scalps in pioneer days. If the natives had used more goose grease and gravy on their hair in the days of their youth, there would not be so many bald-headed men walking around the Square wearing hats to hide their bare pates.

The next Paris fashions that came into the midst of the quiet, peaceful pioneers of the gay seventies were for the female sex. They were the Grecian bend and the Carolina. The pioneers' name for the Grecian bend was bustle, and for the Carolina, hoop-skirt. The store bustle was made of wire. The pioneers of the female sex, both old and young, took to the new fashions like ducks to water. Those that could not afford store bustles made them out of canvas or factory cotton and stuffed them with feathers, wool, shavings or other light material. They were made in three sizes, large, medium and small, according to the size of the wearer. Oh, I forgot to mention where the ladies wore them. Well, they wore them on their east end when they were travelling west.

The store hoop-skirts were made of wire covered with

cotton of different colors. The very genteel ones were covered with colored silks. The pioneer maiden that could not afford store ones manufactured hers out of willow rods bent around in circles and left to dry in the sun. Some were made of ash strips in the same manner as barrel hoops are made. Some of these hoops were so large they could hardly get them through an ordinary door. Like the bustles, they were made in three sizes, large, medium and small, to suit the wearer. In the days when the hoop-skirt was in all its glory you could not get near your partner to "waltz her round again, Willie." It was done with the tips of the fingers. Neither could you get near your partner in a square dance to swing her around. You had to be content with holding her hand and "sashay" around her. However, the old hoop-skirt had one redeeming feature, if you could not get near your partner yourself to give her a hug or swing in the dance, the other fellow could not either, so your best girl was always safe from the other fellow's hug. No bunny-hugging in the days of the bustle and the hoop-skirt, as your best girl was always well protected, fore and aft, if she wore these fashions.

It took some practice and it was a fine art then to sit down in a chair without the lady showing her ankle. If she happened to show her garter, which was tied below the knee, that was horrible and would almost shock the fiddler, who was always supposed to be a little hard-boiled. I do not think bustles and hoop-skirts would work as well in these days of shorties, panties and nothings as they did in the pioneer days of the gay seventies.

Now, if you readers think the gay nineties have anything on the gay seventies and eighties, spiel it out.

DUNGANNON'S LOVELY GIRLS

Air – The Armagh Boys
(Respectfully inscribed to Mrs. John Treleaven,
the mother of a first-class musical family.)
By Joseph Williamson

You muses fine, that grace the Nine,
Pray aid me to impart
These rambling lines; I feel inclined
To soothe my drooping heart.
I've roamed this country many years,
Ofttimes I've been in per'ls,
But my heart still cheers when I draw near
Dungannon's lovely girls.

There's not one lass that I would pass
From Bridge-end to Lucknow,
I'd meet bright smiles around the Nile,
Likewise on Buffer's Row.
These fairy queens, fresh in their teens,
Might win lords, dukes or earls,
Sweet nymphs so rare, beyond compare,
Dungannon's lovely girls.

In the Diggin's, too, there are a few
Well worthy of renown,
Besides a score, perhaps there's more,
In this our thriving town.
Eyes black and blue, of brightest hue,
With teeth like rows of pearls,
'Twould wound my heart, were I to part,
Dungannon's lovely girls.

Around Belfast, "not least, though last,"
There dwell some lovely dames
With aspects bright, hearts leal and light,
They set men's minds aflame.
In festive halls, at New Year's balls
As round the floor they twirl,
I, too, would prance through the mazy dance
With Belfast's charming girls.

One comely dame, I dare not name,
I'd like to charm to rest;
Great Shakespeare says that music's lays
Can soothe the "savage" breast;
Her witching een and graceful mien
Would melt the heart of churls –
Long may her lyre adorn your choir,
Dungannon's lovely girls.

There is one more I could adore
For her sweet vocal strains,
Near Huron's shore where billows roar
Not far from Maitland's streams;
Complexion fair, bright auburn hair
Of nature's choicest curls –
Full well I ween she should be queen
Amongst thy merry girls.

Now to conclude, may all that's good
Protect the lovely fair;
Like birds in spring that sweetly sing
With mates may they soon pair;
Enjoy their loves like turtledoves
While time life's chariot whirls,
Nor forget the time in youthful prime
We called them lovely girls.

When life is o'er and time's no more
May we all meet above,
In that great throng whose "endless song"
Is still redeeming love;
On golden streets each other greet
'Midst gates of crystal pearls,
To that grand choir may you aspire,
Dungannon's lovely girls.

THE BATTLE ON THE FLATS

This is Joe Williamson's poem on the sham battle of
June 22nd, 1871, composed on the ground and respectfully
inscribed to Major Coleman of the 33rd Huron Battalion.
The air is "Waterloo."

On the 22nd of June, brave boys,
On Maitland Heights we lay.
The bugles shrill the air did fill
On that auspicious day.
When formed in line, the troops looked fine,
And bands did sweetly play.
The Wellingtons, like Britain's sons,
Stood forth in bright array.

The London's, too, and Waterloo
In martial blue did spread.
'Twere hard to say which were that day
Most gallant-like or dread.
The Perths and Bruce (no flag of truce,
No, never, shall display)
Stood ready then as loyal men

Whose hearts knew no dismay.
The Thirty-Third we have deferred
For reason that's well known.
In years no past, none them surpassed,
And still they stand ALONE.
When gallant Ross, on foot or horse,
Gives out the watched – for word,
His daring men stand ready then
To grasp each gun or sword.

The London Field Artillery
Brave Shanly did command.
Their roaring guns make Britain's sons
To join them hand in hand.
For weal or woe they'll meet the foe
Whenever duty calls,
And stand or fall, where'er they go,
Like the men of Derry's walls.

The battle raged, the men engaged,
All ardent, brave and true,
Each man full bent to represent
The Scarlet and the Blue.
They fired away, that glorious day,
And their hearts more dauntless grew,
As they fought once more like sires of yore
On far-famed Waterloo.

Now, lest we might your patience tire,
We'll hasten to a close.
May volunteers aye stand the fire,
And vanquish all their foes.
Should Fenians desecrate our shore
We'll give them British steel,

And teach them that we can once more
Protect our country's weal.

God Save the Queen.

Reminiscence and Lament by the Old Town Clock

I AM the old Town Clock. I have ticked out the minutes and struck the hours for over half a century through rain and shine. When I was young and in my prime you used to keep my face painted and my hands and numbers gold-plated, and I felt proud to look down on the old town; but now, as I am getting old, you neglect me. A few years ago a kind-hearted alderman took pity upon me and put an electric light before my face, so that you could see me at night. That put new life in me for a few years more. Then you took away the light and left me in the dark.

But that is not the worst of my sorrows. You let the trees grow up so high that they hide my face from the Square.

Have I not served you faithfully from my youth up to my old age? Have I not taught your children to tell time? Have they not looked upon my face on their way to school? And they always could tell whether to hurry or slow down, for they knew I always told the truth. And the factory men, the bus-drivers, mail-carriers, hotel men, policemen, and the boys on the corners, all knew I was correct as I tolled off the hours by the ringing of my bell. Have I not called you from your nice warm bed to go to a fire or some other calamity, such as is bound to be brought upon the town by the wheels of fate? But I am not used for this task any more; my bell has to remain silent, and I have to listen to a creaky, whistling siren on the old Town Hall to call the firemen. This makes me feel sad.

But this is not what grieves me most. As I have already said, you have let the trees grow so high that I cannot see what goes on around the Square or in the park. When I first took up my abode on the Court House I could look all around the Square and over the town, and the natives from Colborne and Goderich townships could see my smiling face and tell the time from afar off. But now you cannot see my face to tell the time unless you climb to the top of McLean's block or the Bank of Montreal if you are travelling around the Square, and if you have business at the Town Hall on East street with the town clerk or the tax collector, you will have to travel east as far as Knox church before you will be able to see my face over the tree tops.

I have heard many horrid and shameful remarks about myself, such as, "Why don't the council paint the face of the old clock, so we can see the time, or put the old bundle of works higher up on the Court House, so that we can see the time over the tree tops, or get a new clock? The old fellow must be nearly worn out."

All this makes me sad, and I feel like hiding my face behind my hands. I feel very sorry to think that the rising generation don't want the old Town Clock. I hope when you get tired of me and take me down from the Court House, where I have weathered the storms for years and struck the old years out and the new years in with my old bell, and witnessed much of your joys and sorrows since I came among you – I hope, I say, you will give me to the Historical Society of Goderich or Huron County. Do not sell me to Mike Kennedy, the junk man. Please keep me in the old town, where I have seen you grow up, get married and raise a family. I have seen you spooning in the park. I have seen you marching around the Square, arm in arm with your sweetheart. I have seen you drive around the Square after you were married, and many other happy events in your journey through life. I have also seen your funeral cortege as it passed around the Square on its way to Maitland cemetery. But such is life.

I have seen the policeman on his beat. I have also seen him asleep in the park at night, with no one to watch him but the old Town Clock.

When I first came to town there were salt wells, many cooper shops, hotels, mills, factories, ships and sailors, and fishermen in scores. The first Chief of Police I watched was Sam Reid, a big Irishman. Next was Thomas Sturdy, a native of Goderich; then John Yule, also Alf. Dickson, John Sands, and now your present Chief, Dick Postelethwaite. He likes me well enough, but not well enough to trim the trees to let me see what goes on in the park and around the Square. When I first came to town there was a fence around the park to keep the cows out, and a board sidewalk around the Square. I remember seeing two of the native boys from Colborne township, Dad Bogie and Case Allen, drive their horse and buggy around the Square on this board sidewalk (of course on high gear).

I have seen many horse races around the Square in my youthful days. I remember seeing a foot race, once around the Square, between Jock Adams, of Lucknow, and A. M. Polly, of town, for a bet of $50. It was a close race and A. M. Polly won; he was a younger man than Jock. They ran rival passenger coaches between Goderich and Lucknow in the seventies and eighties, and their rivalry was the cause of the foot race. I have seen, before the trees grew up, many a foot race around the Square. I remember seeing Allen Potter from Benmiller beating all-comers in a foot race, twice around the Square, on a First of July. I also saw Jack Platt win a bicycle race, so many times around the Square, on the old high-wheeled bicycles on the same First of July. In those early days, before the trees were planted in the park, they held the athletic sports in the Court House Park. I have seen Angie Matheson, of town, and the late Thomas McKenzie complete in the heavy sports, throwing the hammer, putting the shot and tossing the caber, and other games. They were the two outstanding athletes at the sports of those days.

As I looked down from the top of the Court House into the park one Twelfth of July at an Orange celebration I saw Mrs. Peter Shea knock down a man with an umbrella. His name was Jim Loutit, and he was a butcher. This was for something he had said to Peter, her husband. Peter was a frail little man. Mrs. Shea was Irish; she took up the challenge and won. (Now, the old Town Clock is telling the truth, for I saw this incident myself when I was a boy. – G.H.G.).

I have seen many celebrated characters pass around the old Square since I have been time-keeper for the old town: Donald Dinnie and Duncan C. Ross, the two Scottish athletes who travelled from Scotland to the Caledonian games held at Lucknow for many years; Lord Dufferin, Governor-General of Canada in the seventies; Sir. John A. MacDonald, Sir Wilfrid Laurier, Sir Charles Tupper, Sir Richard Cartwright, and many other prominent men of those times.

I have seen the two old-time editors of the town papers, Daniel McGillicuddy, of The Signal, and James Mitchell, of The Star, pass by one another on the old Square with fire in their eyes.

Many a familiar figure I miss from the old Square since I began to tick out the time. I will mention a few that I miss very much as I look down on the Square and the path: Jonathan Miller, A. M. Polley, Geo. Graham, Robert McLean, William Campbell, William Postelethwaite, John Knox, Daniel Macdonald, Dr. Holmes, Rev. Dr. Ure, Charles Nairn, Dick Black, and Miss Eloise Skimings (The poetess of Lake Huron). Many others I miss that I would like to mention, and I honor all their memories with a sweet sadness that takes me back to the time when I ticked off the last minutes of their earthly life.

But I am only the old Town Clock, though my wheels are O.K., my bearings are good, and my weights are all right, and so is my wiggle-waggle. If you would only put a fresh coat of paint on my face and gold-leaf my hands and figures, and trim the tree tops, so that the little boys and girls could see my face and not be late for school!

Now, there is Charlie Lee, the mayor; Johnnie Craigie, the reeve; Charlie Humber, the alderman; and Lin Knox, the town clerk. I have seen them all grow up from their childhood days. In their youth they used to look up at my face, so I could tell them when to go home. But since they have all grown up and got high-hatted, they never look at poor me any more. Perhaps they cannot see me for the tree tops or for the dirt on my face and hands; but they are all very nice boys – yes, very nice boys – and all natives of the old town. And I claim to be a native of the old town myself.

And now, boys, as we all grew up together, please do me a favor. Trim the trees, so I can see the boys and girls go around the Square and the policeman standing on the corner, as I saw them in the days of my youth; or raise me higher, so I can see what is going on in the old town and you can see my smiling face again as in the days of yore, and put gold-leaf on my hands and numbers, and paint my face a nice black.

I ask you, Mr. Lin Knox, to see that the council of 1933 does this for the old Town Clock.

As time and tide wait for no man, I must keep going; for that is what you put me up on this Court House for. So I must tick off the minutes and strike the hours as I did in the days of my youth, when everyone could see my smiling face, as well as hear me strike off the hours.

And as you boys, as well as I, are all travelling down the hill of time together, when I strike the last hour, and you look upon my face for the last time, may you have sweet memories of me, for I am just the Old Town Clock.

July, 1933

The Governor-General Comes to Town and Is Shown Around by One of the "Natives" – Mayor for Ninety Minutes

YOUR Excellency the Governor: –
"As Mayor of the old town of Goderich it gives me great pleasure, on behalf of the citizens, to welcome you.

"Now, what do you think of our line-up of band boys, Scouts and soldier boys? – the soldier boys with the medals on their breasts, who won in the World War, you know, Governor (I will just call you Governor; sounds more sociable-like). There go the band boys, playing 'Cock o' the North.' The old town is quite Scotchy, don't you know. I am so glad you called on us, Governor. I have travelled with processions led by the band, but never with a bodyguard of soldiers and a Governor by my side. I will point out places of interest and landmarks as the procession moves along.

"This is East street. That building to the left is the old knitting factory. As our young girls, their mothers and old ladies have discarded wool and cashmere for silk hose, it has put our knitting factory out of business. But, Governor, I like to see the silk hose; it is nearer to nature-like than the old wool hose. That building on the right is the Goderich Organ Factory. In the last forty years it has shipped its organs and other products to all parts of the world, and is still keeping Goderich on the map. That factory on the left is where they manufacture the 'Good Roads Machinery.' It has shipped its machinery to many different places in North and South America, and that machine over there is billed for Africa. It is going strong, this factory."

The Governor: "That is an imposing building we are now coming to, Mr. Mayor."

The Mayor: "Oh, yes, Governor, that is Knox Presbyterian church, and it is the oldest church in the town. Sometimes it is called 'Oatmeal church.' By the way, Governor, that is Reeve Johnnie Craigie's, Alderman David Sproul's and Town Clerk Knox's church also. Look to the left, Governor, and see that steeple over the house-tops; that is the Victoria Street United church. That is Alderman Seabrook's and Chief Postelethwaite's church."

"Alderman Humber is a water man and patronizes the Baptist church. It is situated off the Square across from the public library. I will point it out to you. Alderman Gould, Alderman Worsell and Town Assessor Robertson are all old-time Methodists and patronize North Street United church. Deputy-Reeve Moser being a good Catholic attends St. Peter's church. Alderman Brown and Mayor Lee both patronize the Anglican church, same as yourself, Governor. So you see you are riding around the old town in good company, as far as religion goes. That building to the right is the old town hall; that is where the Town's Council meets to do things – scrap and call each other pet names."

The Governor: "Mr. Mayor, what is that imposing building ahead in the bush?"

The Mayor: "Why, Governor, that is not a bush; that is a park, and the building is the County Court House."

The Governor: "What an imposing dome the Court House has!"

The Mayor: "Oh, yes, Governor, that is the old town clock,

but you cannot see the hands or the figures to tell the time. Depression and economy, you know, go hand-in-hand. The Council cannot afford to paint the hands and figures of the old clock. It would cost twelve dollars and fifty cents to paint the hands and face of the old clock, and we must economize; so you see the old town clock is suffering from the depression. But, Governor, if you cannot see the time on the clock we will stop and let you hear her strike as we return from the dock. It has a lovely tone, this old town clock has. But not much use paying twelve dollars and fifty cents to paint the old clock's face, as all the aldermen and town employees wear wrist watches – all excepting Deputy Reeve Moser, and Jake doesn't need one, as he never stops until the work is done. Anyway, the caretaker of the park and the other corporation employees don't need a clock to stop them. So you see, Governor, when the depression is over and a new mayor and aldermen are elected, they may spend twelve dollars and fifty cents of the people's money to paint the face and hands of the old town clock.

"Now, this is the Square. That building to the right is the Royal Bank. Nice little bank. Sometimes we deposit a little cash with them, just to be sociable-like. Now, that building with iron shutters around the corner on North Street is where all county deeds are registered. On North Street are the North Street United church, McKay Hall, the Anglican church, the old Central school, and St. Peter's church; and if we keep going we will land in the county jail. Nice old jail. See the fine stone wall around it. Built in 1840. Not patronized as it should be by the town's people, but its gates always open to strangers.

"The large building over the way is the Alexandra Marine Hospital, former residence of the late M. C. Cameron, one-time Lieutenant-Governor of the North West Territories. This building has been enlarged and remodelled into a hospital. The old town is very proud of her hospital. Its doors are always open to the sick and afflicted of all races and creeds.

"The Goderich Signal, the Grit paper, is also located on this great North Street thoroughfare. Now, Governor, if the Grits were in power at Ottawa, W. H. Robertson's paper would give you a write-up. But you need not look for any cream puffs from The Goderich Signal, Governor. That building on the corner is the Canadian Bank of Commerce. Occasionally we patronize it, but being a corporation and borrowing a lot of money and sometimes overdrawing our account, it sounds more classy and dignified to read in the town papers, 'The Town Council has borrowed $3,000 from the Bank of Montreal.' Sounds like big business. To the left, Governor, are our two leading hotels. The British was built before I was born. It has had many proprietors in its time. It is well run now, right up-to-date, with running water in every room. The one across the corner is the Bedford, with its spacious dining hall, beautifully decorated, where the Lions' Club and other organizations hold their banquets. Now, this is West Street, to the harbor. That building to the left is the Masonic Hall. The other town paper is printed in that building. It is a good old Tory paper, The Goderich Star, and they have a reporter who gives you a good write-up, with all the frills. I will send you one of the papers, Governor."

The Governor: "That is an imposing structure to the left."

The Mayor: "You are right, Governor. That is our post office. Sir John A. MacDonald, away back in the eighties, gave that post office to the old town for voting Tory and sending Robert Porter to Ottawa instead of M. C. Cameron, Grit. Wasn't that nice of Sir John A.?

"That large red building to the right is the old skating rink. It is one of the old town's assets. By the way, Governor, I might mention the old town has several assets. But, Governor, a town without a few assets would be like a Stotten bottle without any Stotten in it. That residence on the right – quite English, you know – is G. L. Parsons'. By the way, he is manager of the Goderich Elevator Company. That vine-clad cottage on the left is

County Crown Attorney Dudley Holmes' residence. That building to the right as we descend the hill is the old Park House, built by the Canada Company. It is now run as a tourist hotel. Quite classy! To the left is the C.P.R. depot. This is the Western Canada Flour Mills, with a capacity of 2500 barrels per day. Ships flour to all parts of the United Kingdom and to other countries where they eat white bread. They also manufacture salt. To the right of the Mills are the Goderich Elevator and Transit Company's elevators. They tranship millions of bushels of grain every year, as this is the shortest lake and rail route to the sea. On the left is the mineral spring, with a capacity of 1000 barrels per day. It comes from 1200 feet under the ground, clear as crystal. It is remarkable for its medicinal qualities. People who are on the water wagon come from far and near to drink its healing waters. It is all free – nature's gift to men and women who are on the water wagon. Beyond the mineral spring is the bathing beach. Sorry, Governor, it is not the good old summer time, so that you could have seen some of our bathing girls.

"Governor, I wish you to notice that little island in the harbor forninst Parsons' elevators. Your Government claims that island and so does Bill Forrest. Your Government made Bill move to the mainland and took part of the island away and dumped it into the lake. Bill wants $100,000 from your Government for making him move to the mainland and taking half of his island away. But your Government doesn't want to pay Bill $100,000. Now, Governor, if you would just O.K. Bill's little account. You don't know Bill, of course, or you would do it. But you do know George Spotton, who looks after our interests down at Ottawa. Well, George is a heavyweight, but he is just a little David alongside this Goliath, Bill. Governor, if you do not O.K. this account, you will know him when he goes down to Ottawa to collect. The only delegation Bill takes with him when he goes to the capital after his money is his big walking stick and his big moustache; and the words Bill can coin to tell your Government why they should pay him that $100,000 are not found in Webster's dictionary nor the King James version of the Bible. And now, Mr. Governor, as I am a friend of your Government, I ask you to O.K. Bill's account, and your visit to our town will not have been in vain, as Bill is a good sport and never banks his money. He will spend it in town and it will kind o'help us over the depression.

"Here we are, Governor, back to the top of the hill. Right from this bank you can view the evening sunset – one of the most wonderful and beautiful sunsets that nature's God can provide for the eye to behold. And more than that, Governor, you can see the sunset twice in the same day. War illusions? Yes, Governor, war illusions. You just go down to the beach and watch the sun sink behind the waters of Lake Huron. Return quickly to the top of the hill and you will see the sun set again behind the waters of old Lake Huron. Now, Governor, this is the Rosedale of Goderich. His Worship Mayor Lee and Reeve Johnnie Craigie and Charley Meakins live in this Rosedale, as well as quite a number of the aristocrats. This is the Hotel Sunset, with 100 rooms with running water and private baths. A nice place, Governor, to spend a few weeks in summer, away from the business worries of Ottawa. This is Britannia Road, and this the Collegiate. Three hundred pupils attend at present. This is Victoria school."

The Governor: "The imposing structure and the grounds seem to be beautifully kept."

The Mayor: "See, Governor, those four smokestacks in the distance, like the Ford plat at Walkerville. Well, that is the North American Chemical Salt Works. Mr. Wurtele is manager."

The Governor: "Oh, yes, Mr. Mayor, I remember your introducing me to Mr. Wurtele. Rather an aristocratic gentleman; very clever, I presume."

The Mayor: "Yes, you are right, Governor. Mr. Wurtele is all you say, with the frills. He manufactures all kinds of salt – table, dairy, package, field, rock – and ships to all parts of the world where they need salt. I have no doubt you use Mr. Wurtele's salt on your table at Rideau Hall. If not, I am sure Mr. Wurtele would send you a bag; and when you get back to Ottawa and your parliament meets again, and you cannot keep the members from kicking over the trace chains, especially the Grit members, if they get too fresh just try some of Mr. Wurtele's Goderich salt on them.

"Here we are back to the depot. Time is up so good-bye, Governor. Remember me to King George and family. Hope you have enjoyed your visit. Come again in the good old summer time; and as you travel to and fro upon God's green earth, if you can, put in a word for the old town of Goderich on the banks of Lake Huron, where you can see the sun set twice in one day.

"And don't forget to O.K. Bill Forrest's account."

October, 1933

The Old Town Clock's Welcome to the Old Boys and Girls

I AM so glad you are coming, as I hear it whispered by the Town Fathers that they are going to paint my face and hands and doll me up for the occasion. It makes me feel as if I were young again, when I was a charming brunette with gold hands and figures. How I long for those days again.

For the last three years I have been very sorrowful. At times I have refused to go. Some of the Town Fathers who had a touch of Hollywood and Mae West in their blood thought it would be wise and dignified to change me from a brunette to a blonde – a real platinum blonde. So a can of aluminum paint for my face, and a quarter-pound of lamp black for my hands and figures. When the painters got through decorating me – grape nuts and shredded wheat! – I looked as if I had drunk a gallon of sour beer, and I'll be d—d if I didn't feel that way, too. So why should I not be tickled to welcome you all to this reunion, as this may be the last time I shall be dolled up for such an occasion.

As you know, I am like yourselves growing old, and the trees in the park are growing so high that you cannot see my face and I cannot see what is going on around the Square as I used to. So about all the good I am now is to strike off the hours. Something seems to whisper to me that before another old boys' and girls' reunion I shall be scrapped. If this should be so, I hope the Town Fathers will put me in the historical museum, as a relic of those bygone days of the last century when you could buy an eight-gallon keg of Henry Wells' beer for $1.25, a cord of four-foot maple wood for $2.50, and a hundred-weight of farmers' dressed pealed pork for $3.00.

However, you will not see much change in the old town since you visited her ten years ago, when she was one hundred years old and celebrated her Centennial. You will find the same old Town Hall (touched up a bit, inside and out), the same Mayor, the same Town Clerk, the same Chief of Police, the same Fire Chief, and the same old water-wagon – but all of them, like yourselves, ten years older. They will all welcome you and help you have a good time. Don't be afraid, Old Boys and Girls, of being run over by the old water-wagon, as the Town Fathers have locked it up for Old Home Week. The old Bedford and British Exchange hotels have both changed proprietors and dressed up in new suits. You will still find the three banks on the Square, but with new managers. If you have more cash than you wish to squander in Old Home Week you can deposit the surplus in any of the three banks, or if you run shy they will lend you a few shekels to go home on – if you are worth the powder. You will find the Goderich Salt Works going strong. If you feel too fresh, get a bag of Wurtele's table salt; it will keep you prime. You will find the town has an up-to-date hospital, with nice matrons and lovely nurses, if you should have occasion to visit it. You will find the jail in the same old place; and if you wish to play golf you will find down on the flats below the jail a fine golf course where the "pro," Jack Annat, from Montrose, Scotland, will take you around for a dollar. If you wish to get away from the hullaballoo around the Square and Hamilton street and Victoria Park, try the Rosedale part of the town, down by the Park House and Charlie Lee's Hotel Sunset, for a quiet time among the aristocrats. When you are rested, go down the Harbor Hill and see the big plant of the

Western Canada Flour Mills Company, the immense elevators of the Goderich Elevator Company; also Bill Forrest and his island. The island is cut down quite a bit; but Bill is as big and jovial as ever. If you don't like the beer up town, or if you have drunk too much of it, you might like to go over to the flowing mineral spring and have a good medicinal drink. Then on to the bathing beach, where you will see the lovely female form divine without having to make a trip to Grand Bend or Bayfield.

There is one thing that saddens me. That is to see the old Goderich Signal moved to a new home. For fifty years I have watched thousands of people of all walks of life pass in and out of that old Signal office. I have seen editors change, and proprietors change. I have seen the present editor, W. H. Robertson, go in that office door as a cub reporter and walk out again a full-fledged editor and proprietor of the old Signal. If you happen to travel down North Street and pass the old Signal office, and are gifted with the bump of imagination, you may glance in the window and see the editors sitting at their desks, the reporters buzzing about, the printers getting The Signal out with the old hand-press, the printer's devil flying around inking the rollers, and the carrier boys waiting their turn for the papers to be delivered to their town customers as in the days of Thomas McQueen.

But alas! where are they all today? – Passed into history. Do you remember the Signal rooster with the Scotch plaid trousers that used to appear and crow in The Signal after a Grit victory?

Now The Signal is printed under the same roof as The Star. Can any old-time Grit imagine Dan McGillicuddy having his Signal printed under the same roof as Jim Mitchell's Goderich Star? Some day in the future we may see the editors of The Orange Sentinel and The Catholic Record getting together in similar fashion. So mote it be.

But, boys and girls, you will see a great change in the natives of Goderich also. When you were here ten years ago the females of the species wore long skirts, and it was like a penny peep-show

THE FAMOUS SIGNAL ROOSTER used to "appear and crow" on the front page of The Signal after any Grit (Liberal) political victory.

to see the calf of a female leg. Year by year the skirts have got shorter and shorter – but don't think for a moment that the old Town Clock is kicking about that. Did not your ancestors, Adam and Eve, wear shorts? They were driven out of the Garden of Eden, it is true, but not for wearing shorts, and, as it should be, the days of long skirts, dust and microbe catchers, have gone with the winds.

Again a sadness creeps over me as I think of the old boys and girls whose last hours I have struck out with my bell since your last reunion ten years ago. I will mention just a few – William Campbell, oldest of all, Alex. Saunders, Robert McLea, Robert Cutt, Charlie Nairn, Charlie Reid, Billy Murney, Billy Brophey, Walter Saults, Dick Black, "father" of the Old Men's Club, Tom Connon and George Williams, and four doctors, Dr. Taylor, Dr. Emmerson, Dr. Hunter and Dr. Whitely – the last-named a native-born whom I miss very much, as I have watched him drive up to his office and around the Square in all kinds of weather for over fifty years. There are many others that I do not mention, old and young, whom my bell has tolled out, as I will continue to do as long as they leave me here and wind me up.

And now, as I look down upon you from my home on the old Court House in Old Home Week and watch you frolic and dance and play around, and go to school and church as in bygone days, I shall wish you a happy time, a safe return to your homes, and sweet memories of the friends you meet in Goderich in Old Home Week.

THE OLD TOWN CLOCK
July 26, 1937.

The Toll Gate on the Old Gravel Road

AFTER passing through the picturesque town of Goderich, travelling northward along the Blue Water highway, we cross the Maitland River into the pretty little village of Saltford. The iron bridge we cross was built fifty-six years ago. The old Maitland had two wooden bridges at this crossing before the present iron bridge was built. The first bridge was built by the old Canada Company, when this little village was called Slabtown. The next wooden bridge was built, I believe, by the Gravel Road Company that controlled the road from Goderich to Lucknow. This is where the old toll-gates come into the picture - six, I think, between Goderich and Lucknow. Then Slabtown changed its name to Maitlandville. In those days the village could boast of seven salt wells and their cooper shops, two taverns, two stores, two tanneries, a brewery and hop yard, a brickyard, a lime kiln, Henderson the weaver, MacLaren the tailor, Buchanan the shoe-maker, Gallagher the harness-maker, Schultz the cigar-maker, McIntyre the blacksmith and Sandy Donaldson the carpenter, and David Lawson's sawmill on the creek at the foot of the hill; also a school house and a Temperance hall, which still survives – I mean the building, not the members. As I pass it by I have sweet memories of when I rode the goat in that old building, more than fifty years ago. I was initiated into the hidden mysteries of the Sons of Temperance, and the pledges I made and the obligations I took; alas, to be broken! My conductors during my initiation, and who introduced me to the Grand Mogul for the obligations, were two young ladies, one of the brunette type of beauty, with lovely black eyes – Charity Long was her name; the other of the blonde type, with rosy cheeks – Nancy Gilders was her name. There were visitors present from Goderich Lodge, one of them named Frank Elliott. After I was initiated, this Elliott boy sang a song, "There Is a Bedbug on Your Collar, Laddie Daw." As I had never ridden a goat before, and as bedbugs in those days were free traders, they had the right of way on any paper collar. I took it all in as part of the Sons of Temperance initiation. The first part of my initiation, in which I was supported by the two young lady conductors, was like a little bit of heaven upon earth. While travelling around the hall with them, I was wishing the promenade had reached to Holmesville. But when they left me in charge of the Grand Mogul the picture changed. I thought I was in Dante's Inferno, as pictured by Gustav Dore, as I was shown the evils of John Barleycorn, and all his little corns, so that I was afraid of all kinds of liquor except running water, green tea, buttermilk, lemonade, and Carman's ginger beer, which was a temperance drink. I shunned them all. A tavernkeeper, I thought, was doomed to the lake of fire and brimstone, and all his family unto the third and fourth generation, and anyone who went inside of a building with the sign "Beer, Wine and Spirituous Liquors Sold Here" was a sure candidate for hellfire and brimstone. As time marched on, however, I am sorry to say, I fell off the water wagon and like Mulligan's train have been off and on ever since. I will take off my hat to anyone that can keep the Sons of Temperance obligations. One thing I do know, you will escape a lot of drinks flavored with hell fire and brimstone and a sore head the next morning after.

As the salt wells boomed the little village, the natives changed the name again to Saltford, in honor of the salt wells.

As we leave the little village and proceed northward and

come to the foot of Dunlop's Hill, we encounter the old toll-gate No. 1 that bars our way. The toll-gate keeper was one Richard Postelethwaite, grandfather of Richard Postelethwaite, the present Chief of Police of Goderich. Postelethwaite had formerly been employed by the Gravel Road Co. as gate-keeper near Glenn's Hill, between Dungannon and Lucknow. The Company moved him to Saltford, as it was the busiest place and he the ablest gate-keeper. In those days there was a wild lot of boys and big men who came down the Lake Shore road from Sheppardton, Port Albert, Kingsbridge and Kintail, and who sometimes did not think the Goderich Northern Gravel Road Company had any right to make them pay toll to get into Goderich, especially if they had got liquored up on the Lake Shore road before reaching Saltford, as in those days there were nine taverns between Kintail and this toll-gate. So you see the toll-gate keeper had to be a husky piece of humanity to keep the Lake Shore boys in their place and make them pay their toll before the gate would be opened to let them pass through to Goderich. The toll was: Double team of horses or oxen, 10 cents one way, return same day 15 cents; single, 5 cents; man on horseback, 1 penny (2 cents); foot passengers, funerals and weddings free.

After passing through No. 1 gate northward and casting your eyes heavenward you spy on top of the hill the tomb of the two original "Huron Old Boys," Doctor and Captain Dunlop. Then on up the hill to Garbraid, the Dunlop homestead to the right, on to the Crown and Anchor tavern, which stood on the corner of the late James McManus' farm, where the large elm tree stands today. In those days the farm belonged to the Dunlops. Then on one and a-half miles to Millburn, now Dunlop, to John Allen's tavern, where the Northern Gravel road turns eastward to Crackey Robinson's corner, Smith's Hill, north to Nile, Dungannon, Belfast and Lucknow. The County of Huron purchased this road from the Goderich Northern Gravel Road Company in the early seventies – '73, I think. Hence the old toll-gates have gone and the toll-gate keepers have also gone to that land where the Goderich Northern Gravel Road Company dare not collect the toll; but the Gravel Road Company, their toll-gate keepers and the bad boys from the Lake Shore road and the rest of us will all have to pay our toll to St. Peter before he will open Heaven's gate to us.

P.S. – Frank, of the "Bedbug on Your Collar" song, I see is still in the flesh, and he must have kept his temperance obligations and ridden the water-wagon many miles, or he could not hold his present job. You may see him almost any day riding around the Square and down Hamilton street on a spirited saddle horse which he is training for the sweepstake races and jumping. One thing, Frank, you are nearer heaven when you are on that saddle-horse than when you are on the ground, like yours truly.

Following is a copy of one of the share certificates issued by the Northern Gravel Road Company:

THE GODERICH NORTHERN GRAVEL ROAD COMPANY

Incorporated by Act of Parliament

Shares No.s 1001 to 2000, both inclusive.

This is to certify that George Brown, Esquire, or order, is entitled to one thousand shares of twenty dollars each in the stock of the Goderich Northern Gravel Road Company as per issue of scrip in his name in detail bearing date First February, 1859, viz., Numbers 1001, Ten Hundred and One, to Twenty Hundred, 2000, both inclusive, subject to the provisions of the Act of Parliament authorizing the incorporation of the Company and to the by-laws and regulations passed or hereafter to be passed in accordance therewith.

Dated this Tenth day of April, 1861.

J. B. GORDON,   J. MACDONALD,
Secretary and Treasurer.   President.

The Blue Water Highway, Dunlop to Sheppardton, in Pioneer Days

IN pioneer days this four and a-half miles of Lake Shore road was a busy piece of roadway; it is now part of the "Blue Water Highway." The present Dunlop in those days was called Millburn, which had a fine tavern, built and run by a man named Joe Aptigrove, and afterwards bought and run by John Allen, father of Anthony Allen, a later proprietor, also a grist mill and sawmill built and owned by a Mr. Savage, Archie MacDonald's wagon shop and Hod Horton's blacksmith shop.

Half-a-mile northward was Donald Cummings' carpenter shop, where Dunlop school now stands. Cummings had a big homemade windmill that furnished the motive power for planing, sawing and wood-turning. Donald's big windmill was the cause of many cuss and swear words, as it was built close to the road and often scared horses and caused runaways and put several of the native boys and their girls in the ditch.

Across from Cummings was the Williams' farm. The Williams were English aristocrats, who came out in the early days of the Canada Company, of which Williams was an employee. I believe Mrs. Williams was a Hyndman. Their son, Scarlett, was correspondent for The Huron Signal. He gave the local news of Dunlop and Leeburn in the gay eighties and nineties. He sometimes wrote under the name of Joe Mayweed and some of his items were quaint and interesting. But Scarr, as he was called, has passed on years ago to join the pioneers of Dunlop and Leeburn in the other world. May he have sweet rest there, for he was a kind, inoffensive, childlike, lovable man; would not say an unkind word to offend the most humble of God's creatures.

Then on to Leeburn, to the settlement of the Straughans and the Hortons. Leeburn was named by James Straughan after his birthplace in Scotland. He gave the land for two churches on his farm; one a Bible Christian and the other Presbyterian, built by the enterprising Presbyterian clergyman, Rev. Jas. Sieveright. The young pioneer boys and girls called him "Old Hundred," as he always opened his church service with the Hundredth Psalm ("All people that on earth do dwell, sing to the Lord with cheerful voice"). This Presbyterian church was burned down years ago and was rebuilt on the Linklater farm, where it now stands, and is still a shelter in the time of storm to offspring of the pioneer Presbyterians who joined the church union.

William McVittie had a sawmill on the creek on the Straughan farm. Henry Horton had a blacksmith shop and used to manufacture cow-bells. James Straughan also had a blacksmith and machine shop. Last but not least was a Temperance Hall that did duty for many things besides temperance – political meetings, Dominion, Provincial, municipal, all used this Temperance Hall to get rid of their hot air at election times. It was also used as a polling place for all the different elections. I polled my first vote in that old Temperance Hall. I voted for M. C. Cameron. His opponent was Robert Porter. I lost my vote; I think Porter won by about sixteen votes. All I then knew about politics was that there were Grits and Tories, and that I was a Grit and had to vote for Cameron, as I was born a Grit, my father being a "Scotch Bonnet Grit," which meant the same as dyed-in-the-wool Tory. But I guess the dyed-in-the-wool Tories were in power at Ottawa at this time, as one of them was returning officer at this election and his name was Anthony Allen. Wool or no wool, the same Anthony was a good friend to many of the Scotch Bonnett Grits. The old Temperance Hall does duty at present as a driving shed for a tractor on Roy Linklater's farm.

Then on half-a-mile and turn left towards the lake and it will take you down to that old historic place, the Point Farm on the bank of Lake Huron. All that is left now to remind the traveller on the Blue Water Highway of that once busy spot is the old gateposts and the house on the side of the road, now occupied by Mrs. Jos. Cook, which at one time was a tavern called the Point Farm Branch. The original Point Farm Hotel was built by a man named Davis. His wife was a Hawkins, daughter of the original John Hawkins, pioneer of Port Albert and ancestor of all the Hawkins that still live in and around that historic village of Port Albert. Davis was drowned bringing supplies from Goderich in a rowboat one night in a storm. Afterwards his widow married J. J. Wright, that enterprising, dignified, aristocratic little Englishman.

That put the Point Farm on the map of North America. The first Point Farm building was burned down. J. J. built an enormous building, with rooms for 200 guests, drawing-room 87x22 feet, and a dining-room capable of accommodating 300 diners at one sitting. Tower 75 feet high from the top of which a good view of the country was to be had, and a sight of Michigan across the lake when visibility was good. J. J. Wright's Point Farm in its palmiest days was the only summer resort in Western Ontario. He built a telegraph line from the Lake road to the Point and this was tapped in the G.N.W. Telegraph line. He kept a telegraph operator during the tourist season. He also ran a passenger bus between Point Farm and Goderich for the accommodation of guests, meeting arrival and departure of all trains and boats at Goderich. And if you did not wish to travel out from Goderich by bus, you could go by boat, as this enterprising man had a little steam yacht built, the "Tommy Wright," which for a time, in the palmy days of the Point Farm, ran between Goderich harbor and that resort.

Mrs. Chas. Wells of Goderich states that it was her father-in-law, Charles Wells senior, who owned and ran the boat. At that time the Point Farm was during the summer season a busy, active place, the resort of well-to-do people from the United States as well as from points in Canada. Mr. Wells conceived the idea of a passenger boat plying between Goderich harbor and the Point – a distance by water of about four miles – and Mr. Wright falling in with the suggestion Mr. Wells had a steam launch fitted out and he himself navigated it. The present Mrs. Wells remembers it quite well and says it was a very pretty little boat. Besides running to the Point Farm it would sometimes take excursions out on the lake, and sometimes the town band would be taken along to enliven the trip. The proprietor of the Point Farm had a son, Thomas, and in compliment to him Mr. Wells named his boat the "Tommy Wright." One night while lying in the harbor at Goderich it took fire and burned to the water's edge. Mrs. Wells still has in her possession the flag and some other relics of the little craft.

The old Point Farm housed over 150 guests at one time in its hey-day. J. J. was a very patriotic Englishman; had the Union Jack always flying on the tower of the hotel. There was also a flagpole on the lake bank where flew another Union Jack. He had a small brass cannon mounted on the bank which he would fire off to salute all passenger boats that passed up or down the lake. They were many in those days; many of them brought grist to his mill.

Mr. Wright when he left Point Farm presented the brass cannon to R. S. Williams of the Bank of Commerce. If anyone reads this and could trace this little brass cannon old-timers would be pleased to have it donated to the Goderich Historical Society as a pioneer relic of the Point Farm, and the proprietor J. J. Wright.

The 24th of May and July 1st were always big days at Point Farm. Picnickers came from miles around to enjoy a day's outing at the resort, where there were swings, summer houses, lovers' lanes, a good bathing beach, race course, ball grounds. Sunday schools, day schools, temperance lodges and other societies picnicked together when I was a boy, when there must have been over 2,000 people on the grounds. But time wore on and smaller resorts sprang up and old Point Farm, like all great things, had its day. In its prime it was a credit to its proprietor, J. J. Wright, and a great benefit to the country around, for many a dollar the farmers and their sons and daughters, wives and children got from the Point Farm for poultry, vegetables, butter, eggs, berries, milk, cream and fruit; and the daughters as waitresses and maids to tourist parties. In the height of the old Point Farm popularity J. J. had a cottage built at the lake road and a gatekeeper employed to open and shut the gate to its patrons. This was "quite English, you know."

Now this is the saddest part to write, but I tell the truth as near as I can remember. Nothing remains now of the beautiful building, the summer houses, grounds, lovers' lane, etc. The beautiful furniture was sold at auction and the beautiful building sold to wreckers and the thirty acres comprised in the fairy grounds where dainty feet once trod is now a pasture field for the treading of lesser cattle. J. J. himself has gone to meet his patrons in another world. He had one son, Thomas, who lives retired in England.

Continuing a mile farther along the road, where it turns to the lake at Bogie's Beach was called Haley's Corners. A man named Haley, a retired Army sergeant, kept a tavern where Wilfred Smith now lives. Pat Kelly, grandfather of the Foley boys, lived on the corner south. James Strachan, jr., had a saw- mill on Kelly's Creek at this point. Then on a half-mile was Capt. Andrew Bogie's farm where my bundle of joy was born. She is still with me as a joy, but a bigger bundle than she was fifty years ago.

Then on another half-mile to Dick Carney's farm, where Dick Jewell had a steam sawmill on the creek. Then on one mile to Sheppardton, my old home town, which at one time could boast of a tavern, two stores, blacksmith shop, Methodist church,

Orange Hall (L.O.L. 383), and a school, which still stands. In the early days Sheppardton was called the "Holy Cross," so named from the way the Lake Shore road connected with the boundary line road between Colborne and Ashfield. This boundary line stops at Lake Huron about one mile west of Sheppardton and is the northern boundary of the Canada Company lands.

When Sheppardton got a postoffice it got the name Sheppardton, called after a settler named Sheppard. R.T. Haynes was the first postmaster. His wife was one of this settler Sheppard's daughters. None of the original Sheppards are left around here except Ben Sheppard of Goderich. W. G. Hawkins of Sheppardton, Frank Hawkins or Toronto and Mrs. Bessie Hawkins Walden of Colorado Springs, Colo., U.S.A., are grandchildren of the original Sheppard, as their mother was a Sheppard. And as I retrace my steps back from Sheppardton to Dunlop the only original landmarks that are left are the brick cottage at Sheppardton, erected by R. T. Haynes on the old Sheppard estate, and the house on the side of the road which was the Point Farm winter branch, and Leeburn church.

The only names of the original old settlers found among those still living on the Lake Shore road are those of the Bogies, Foleys, Linklaters, Fulfords, Chisholms, Cluttons and Hortons. The rest of the early settlers have passed on like the pioneers and the passenger pigeons to the beautiful Isle of Somewhere.

P.S. – I remember one Saturday evening when Andy McAllister was running the Point Farm tavern on the side of the road, David and I were walking home from Dunlop, where we were working in Hodge's sawmill. As we were boys in our teens we thought we would go into Andy's and get a glass of beer. We each spent 10 cents and had two glasses of beer each. Andy must have put a stick in it. I was carrying a pasteboard biscuit box with a kitten in it, a present for my mother from Mrs. Anthony Allen. Henry Wells' hops and yeast began to work on David's innards. He thought I had carried the kitten far enough, so he up and kicked the box out of my hand; kicked the box to pieces, and the kitten ran up a telegraph pole. I "depokerated" and we had a scrap on the side of the road; both rolled into the ditch, David on top. Billy Morrish came along in a buggy, got out and separated us. The last I saw of the kitten it was sitting on top of the telegraph pole. By the time we reached home we had got the beer pretty well out of our systems. We gave mother our wages, and were commended by her for bringing home our wages like good boys. We said nothing about the beer nor the scrap nor the kitten, until years afterward.

"Depokerated" – I don't know how to spell it, nor do I know its meaning; but I felt like depokerating brother David when he kicked my catbox to pieces and the kitten looked down from the top of a telegraph pole at me. This word "depokerated" was coined by one of the old pioneers of Sheppardton School. Before the days of the school inspector, the parish minister or the most learned of the trustees generally examined the pupils for promotion. Spelling was the high water mark of a good scholar. It did not matter if you could not bound the County of Huron, nor name the county town, say the multiplication tables backwards, or work out the square root by the rule of three. If you were a good speller you were a candidate for promotion. Well, this school trustee came one day to examine the pupils for promotion. Started at fifth class. A boy named Charlie was at the head of the class. "Chas. Hawkins, spell depokerated." "D-E-P-K-O-K———." "Wrong." So were all the rest of the class, which included one of this trustee's own daughters, who afterwards qualified as a school teacher. This word "depokerated" travelled all the way down from fifth class to second without anyone spelling it to suit this trustee examiner. But the day of the pioneer school trustee examiner has passed into history. There has been many a new word coined by college professors since this pioneer school trustee examiner coined "depokerated." If you want to get rid of hot air, and the college professor's new-coined words don't have the kick in them, just try Geo. B.'s word depokerated and you will find it has the Brotherly Love Kick in it, like brandy sauce on plum pudding.

LETTER FROM MR. T.C. WRIGHT

26 Riverside Road,
South Bourne,
Bournemouth, England.

Dear Mr. Green, –

Recently I received two copies of The Signal-Star – one from my old friend Charles Ellis of Goderich; also one from my cousin in Vancouver, J. B. Hawkins, containing your article on the Dunlop-Sheppardton road. Please accept my thanks for the kind references to my dear old father and my boyhood home. I must compliment you on the accuracy of your statements so far as the Point Farm was concerned. You really gave details which I thought I only knew. Mr. Davis was a young Englishman of a very good family. Came to Canada to farm and bought four hundred acres, what was then known as Four Mile Point. Nearly all this land was disposed of by my father, who kept only the big orchard and the grounds around the big house, for hotel requirements – about thirty acres.

There is a story of my mother's first husband, which was told to me by the late Ira Lewis, K.C.. Mr. Lewis and Davis were having dinner with my grandfather Hawkins in Port Albert, where grandfather asked his son-in-law, Davis, to say grace. Davis gazed at the joint on the table, and replied, "I'm damned if I will, for pork."

Now, coming back to the Point Farm days: You may not know that the large barn there was raised by sailors of the Royal Navy, under the following circumstances: My father had as guests at the old Huron Hotel in Goderich, where the Oddfellows' Hall now stands, Captain Huntley and his wife during the Fenian Raid excitement. Captain Huntley commanded Her Majesty's gunboat Cherub, which lay in Goderich for a year or more at that time. As men were scarce, my father asked the Captain to let him have some men to raise a barn. Captain Huntley explained that he could not report to the Admiralty that his men had gone off to raise a barn, but he would be delighted to give twelve good conduct men a day's leave and they could raise a barn or raise the devil, so long as they got back to the ship at a certain time. I imagine this was the only barn in Canada raised by sailors from the Royal Navy.

Goderich and vicinity in the pioneer days had a reputation for hard drinking; when it was said that Goderich men drank hot Scotch whisky in August, before breakfast. I met a man who claimed he did it and said the town was so healthy there were only two diseases of which the natives died – old age and delirium tremens.

As you are doubtless aware, we are having an anxious time over here just now and I would not bet a sixpence on peace or war, but hope and pray for peace as all the churches are doing here. I would feel far safer looking out over Lake Huron than over the English Channel, half-a-mile away from where I am writing this.

And once more thanking you for your interesting articles, with kind regards to Mrs. Green and yourself, believe me,

Very sincerely yours,

T. C. WRIGHT.
October 14, 1938.

N.B. – I might say you can see the two original photograph pictures of the gunboat Cherub, taken as she lay in Goderich harbor during the Fenian Raid days.

I got these two above-mentioned photos from the late J. J. Wright and donated them to the Goderich Historical Society.

G.H.G.

A Huron Old Boy Visits the Old Home Town After an Absence of Fifty Years

I LEFT the old home town by the old Grand Trunk Railway. Jim Miller, engineer; Alf. Saults, fireman; Thomas Ausebrook, conductor; Peter MacFarlane, brakeman; William Tye, mail clerk; Frank Lawrence, express and baggageman. These have all made their last run on the old Grand Trunk and have been sidetracked on to that spur line that has carried them to the land where we trust they will find the run much smoother than it was between Stratford and Goderich in the days of the old wood-burner and hand-brake.

When I returned I came in on the C.P.R. electric-drawn train; landed down at the dock at a fine brick depot, near where big Colin McIvor once lived. I looked around for Captain Babb's Ocean House, but instead saw big oil tanks and cement elevators. I looked for A. M. Polley's or Tom Swarts' buses, but, instead, a dapper young man, whom I took for a school teacher, says to me, "Taxi up town, mister?" "Yes, take me to the old Albion Hotel." "Please, sir, I never heard of it." "Well, is the old British Exchange around?" "Oh, yes." "Well, take me there."

I was soon ushered up to the door, but what a surprise I got when I stepped inside. Since I last stood there in the days when George B. Cox was proprietor – why, I was in a real vestibule and hall reception room, with inlaid marble-tiled floors, leather-covered armchairs, chesterfield lounging suites, and peek-a-boo blinds on the dining-room windows. What a change from the days of the old wooden-arm bar-room chairs! I met a very cheerful and important-looking landlord who I found out was a barefoot boy running around Port Albert, when I last stood inside the old British. His wife, the landlady, is a daughter of Bob Campbell, who was lighthouse-keeper when last I saw the sun set over Lake Huron, and this landlady a little rosy-cheeked school girl.

As I rose next morning, much refreshed, I found there was real Lake Huron water on tap in my room. I wandered out to take a walk around the Square. Ah, what a change! The old Court House looked familiar. The old town clock I could hardly see for trees. In my boyhood days there were no trees in Court House Park, but there was a chain fence around it with a turnstile gate at each crossroad to keep the cows out. The dear old bossie cows. How I missed them! I used to meet them on my way to school. The dear old bossie always had the right of way. She had a good time in the old town when I was a boy – eating the grass off the streets and lawns, also the vegetables in the natives' gardens. But that was all before the lawnmower came to town; then bossie had to go to grass down on the river flats. The old plank sidewalk around the Square – gone; and the board verandas that kept the sun from fading the goods in the store windows – gone. But replaced with cement walks and awnings of many colors like Joseph's coat.

As I came to the site of the old Albion Hotel I found it had been gutted by fire years ago. It had been remodelled and its name changed to "The Bedford," the family name of the owners. Sure I remember Jack Bedford and his sister Lizzie ("Toots" for short). I found the Bedford also had tile marble floors, easy chairs in the reception room, with the latest shutter blinds to keep the passers-by from gawking in at the aristocrats inside. Yes, a real New York Fifth avenue, Waldorf Astor style in the old town. As I

167

walked along I looked for Donald Strachan's grocery store, and Cattle's drug store; across Kingston street to the Crabb block; Christopher Crabb's store, where you could buy anything from a needle to an anchor. At this time Christopher Crabb was Chief Magistrate. His first case was from the township of Ashfield. A neighborly free-for-all-fight. After Christopher heard the evidence he fined each witness one dollar each, the plaintiff and defendant the costs of the court. This pleased both litigants, as they left the court arm in arm, took all their witnesses down to Henry Martin's hotel, gave them their dinners and free drinks all around. This is the only case on record in Huron County where all hands were fined for the upkeep of a Court of Justice and where both plaintiff and defendant and their witnesses all went home with their bellies full of good eats and whiskey. Peace and harmony reigns to this day in that neighborhood.

I looked in vain for Major Cooke's liquor store, Card's tailor shop, E. & J. Downing's boot and shoe store; across East Street Mitchell the tailor, Abraham Smith's tailor shop, the law office of Cameron, Holt & Cameron, Henry Horton's grocery where Charlie Nairn was chief salesman, who afterwards bought out Horton and carried on the business himself for many years. Over this old store the old-time photographers took your tintype – Geo. Robson, Geo. Stewart, Tom Brophey. Those were the days of the iron fork prop to keep your head steady and look pretty while getting your picture taken for your little sweetie; and the female of the species backed her pretty little head into this same iron steady-all to get her tintype taken for her boy friend to carry around in the locket on his watch chain or in his vest pocket over his heart. The old photo gallery is in the same place, but the old photographers, the iron prop, the tintype pictures, the boy friends and the little sweeties are all gone.

When Abraham Smith was carrying on the tailoring business a man from Colborne township named Wilson Olds came in and was measured for a suit of clothes at Smith's. "Abraham, will you take a load of wood on the clothes?" he asked. "Yes, Wilson, if it is good dry wood." When the clothes were finished Wilson brought in the load of wood. He had put a hayrack on his sleigh with two cords of wood; then got two neighbors' teams, with one and a-half cords on each; landed up at Jim Bailey's hotel on Hamilton Street, loaded the five cords on the hayrack sleigh, drove into Abraham's back yard, unloaded and told Abraham to come and measure the wood. Abraham being a peaceful man accepted the wood as a load. He had to give Wilson the suit of clothes and $3.50 cash money to square the deal. This was the biggest load of wood sold in Goderich. This Wilson Olds was what we would call today a high-pressure salesman. When the first rotary hay and straw cutting-box came on the market, manufactured by Thompson and Williams of Mitchell, Wilson was agent for Ashfield and Colborne townships and it has been said he sold a cutting-box to nearly every farmer and to some people who had neither horse, cow, hay or straw. Wilson afterwards went to Dakota, and I believe he lost his life fighting a prairie fire.

To resume my survey of the Square: across Hamilton Street John Storey's old stand, remodelled and now occupied by the Royal Bank of Canada; Colborne Bros.' dry goods store – all gone. The first familiar name to greet my eyes in my walk around the Square was that of W. Acheson & Son. I stopped, went inside and found a tall, distinguished-looking gentleman with silvery hair, who told me he was Jack Acheson. I remember him as a boy, also his pretty little sisters when I left the old town; their father was William Acheson, harness-maker, on Hamilton Street. There were also John and George Acheson, dry goods merchants. Then the McLeans – another familiar name. Allan P. McLean I remember very well. He wore kilts on special occasions, and in cold weather a plaid and Scotch bonnet. His two sons Allan and William still carry on the tailoring and clothing business. Jimmy Wilson, druggist, gone; but the drug store was there with the name E. R. Wigle over the door. W. T. Welsh, watchmaker and jeweller, gone. C. G. Newton, the hatter – in his old stand was a familiar name, Fred Price, grocer. Yes, I remember his grandfather, Rees Price, and his father St. George Price. I went in and shook St. George by the hand. We talked about our boyhood days at school and the jolly times we had around the dock. When I left Goderich, his father, Rees Price, kept a grocery store where the C.P.R. ticket office now stands.

Crossed North Street. Sam Detlor's clothing emporium, gone (now and A. & P. store); Horace Horton's private bank, gone; Moorhouse's, later Imrie's, and still later Fraser & Porter's bookstore. McKenzie & Wilkinson's hardware store; Barney McCormick's and Paddy O'Dea's clothing stores, Croft & Johnston's dry goods store, R. B. Smith's dry goods store, and Frank Jordan's drug store – all gone.

Across Colborne Street – Ball's restaurant, where we got a dish of pure ice cream for five cents, gone, but in its place the Canadian Bank of Commerce stands out boldly inviting you in, but about all the cram you get there for a nickel is a pleasant smile. O. G. Martin, watch maker; Bunty Munro, haberdasher; Bill Campbell's boot and shoe store, Davis' tin shop, Geo. Thomson's music store. Geo. Grant the grocer – all gone. Another familiar name: Pridham the tailor still does business in high-class tailoring. The Geo. Acheson opera house on the corner, gone, but in its place the old aristocratic Bank of Montreal. I remember attending a Grit political meeting in this old opera house. There was present a noisy Tory from Goderich township who kept interrupting the speaker (M. C. Cameron). A. M. Polley grabbed this heckler by the neck and the seat of the pants and threw him downstairs, then hollered, "Go ahead, Cameron, Goderich township has gone out the door."

Across West Street, John Kay, dry goods; Rees Price, grocery; Grand Trunk telegraph and ticket office, Harry Armstrong agent; Curry Bros.' Huron Hotel, Bingham's restaurant, Andrews Bros., butchers; George McIntosh, grocer; E. L. Johnston and R. R. Sallows the photographers, Williams the barber; Jim Moss and

Tom Hall, shoe makers, and John Butler's book store – all gone. When John Butler was chief magistrate, there was a man that did not have a Sunday school teacher's certificate for good conduct and who was brought before the magistrate,, charged with stealing a shirt out of one of the stores on the Square. Magistrate Butler says to him, "Prisoner in the box, stand up; are you guilty or not guilty?" "Not guilty, Mr. Butler." "You lie, you buggar, I see the shirt on you now. Ten days in jail with hard labor." This poor fellow's reputation got to court ahead of himself. Another familiar name caught my eye – McLean's butcher shop. When I left town there were four McLean brothers in town, cattle-buyers and butchers. These four brothers are all gone, but their sons carry on business in the McLean blocks.

As I made the circle of the Square I saw only five familiar names that did business there when I was a boy. I might say right here that, in all my travels, I never came across a town where the Square was round, as in my old home town. I felt a lonesomeness creep over me, but when the old town clock struck the hour of ten bells it cheered me up and carried me back to my boyhood days when I used to hear the bell ring out to warn the citizens of a fire. But what I seem to miss most of all is the old pioneer teamsters. I miss Jim Doyle's express wagon, Charlie Washington's and Freddie Platt's two-horse drays. Then there were the one-horse drays, Hugh Munro's, Dannie Campbell's, Bill Postelethwaite's and Edward Hopper's. Then the old teamsters that drew the barrels of salt from the wells to the dock, also gravel, stone, lumber, wood and hay: Dave Reid, Sanford Stokes, Hugh McGrattan, John McEvoy, Pat McCarthy, Peter O'Rourke, Bob Duff, Jack Sands, John McKinnon, Jimmy Jones, Jack Barker and John Beacom. I knew them all, as I got many a ride with them when I was a boy.

I missed A. M. Polley's carriage team of "spots" that conveyed wedding parties and other dignitaries around the Square; and H. Y. Attrill's spirited carriage team with his darky coachman from the Attrill estate, across the river (now the Fleming estate), and M. C. Cameron's team of dark bay's that drew the Grit members and imported speakers around the Square at election times, and many other dignitaries, including Lord Dufferin when he visited the old town as Governor-General in 1874. At one time David Munro, still living in the old town, was engaged at the age of fifteen as coachman for M. C. Cameron and remained with him until he was eighteen. As a boy I used to envy Dave his job, as he looked so dignified and important all dressed up and sitting up on the high seat driving Cameron's carriage with its occupants around the Square, and the good eats and drinks Dave got. I missed also the old band boys, Dick Parker, Ben and Tom Armstrong, Bob and Dave Black, Frank and Wilmer Smith, Bennie, Jim and Josh Thomas, Jack Story, Jim Wells, the Donaghs and others I cannot now recall.

I miss the old original natives off the streets: such as Dan Moran, the priest's man, who rang the bell in the old Catholic church, Mike Crae, Mike O'Reilly, Peter and Mrs. Shea, their sons Peter and Andrew, Dan Graham, Henry Sillib the woodturner, John Murray, Dan McLeod, Miss Eloise Skimings, and her brother William, Pat Nugent and many others. And as I walked through the Court House I missed Sheriff Gibbons, Bob Reynolds, Judge Toms, Judge Doyle, Ira Lewis, Dr. Holmes, Dan McDonald, William Lane, William McCreath; and at the Town Hall, Wm. Mitchell, and the policemen, Sam Reid, Black Tom Sturdy, John Yule. Then there were the old auctioneers, Trueman, J. C. Currie, and Jack Knox, whom I liked to hear spiel at auction sales, and the old British soldier, Tom Huckstep, who acted as town bill-poster; and the old Doctors, McDougall, Shannon, McLean, McMicking, McLeod, Taylor, Cassidy, and Dr. Nicholson, dentist. I remember him, as he once pulled a tooth for me. I thought my head came off. No freezing gums nor laughing gas in those days: just sit in chair and hang on with both hands, dentist grabs tooth with big forceps that fill your mouth, hangs on and yanks you around the room until tooth comes out. But all of this cost only a quarter.

I remember the old school teachers at Central – W. R. Miller, principal, Mr. Annis (one arm), Miss Trainor, Miss Longworth, Miss Norval, Miss Dickson and Miss Bond. At Grammar School, Hugh I. Strang, A. J. Moore and S. P. Halls are all I recall. The old Catholic church – Fathers Waters and West; St. George's, Rev. Archdeacon Elwood; North Street Methodist, Revs. T. M. Campbell and G. R. Turk; old Gaelic church, Revs. Sieveright and Fletcher. Then the old editors of the home town papers, Dan McGillicuddy of The Signal and James Mitchell of The Star, who threw fire and brimstone at each other through the editorials in their papers. I trust that if they meet in the other world they will lick thumbs and shun editorials that smell of brimstone. Almost forgot old Knox church, Rev. Dr. Ure and J. A. Turnbull. It was in this church, as a boy, I saw and heard the first pipe-organ. It was played by a girl named Nettie Seegmiller. She was very fair and had nice rosy cheeks, and as I looked upon her and heard her play, the music seemed so sweet I thought she was an angel sent down from heaven. "It was a childish ignorance, but now 'tis little joy to know I'm farther off from heaven than when I was a boy."

As I continued my walk I came to the Maitland cemetery and as I wandered amongst the tombs I noticed many names I recalled; many I had almost forgotten. Some of the names carved upon the monuments caused my thoughts to wander back to my boyhood days, as I shed a silent tear for sweet memories of bygone days, and as I turned my steps back to the old town my mind wandered to my last earthly home. I could not wish it to be in a more beautiful spot than the Maitland cemetery, that beautiful home of the dead where sleep ten thousand of the pioneers of my old home town. And as I viewed the scenes of the old town and noted the many changes and the new faces I met in my travels around the Square, and memory brought back the faces of those that I missed, my prayer was: May we one and all have sweet rest in the old Home Town beyond the clouds.

Biographical Profile for
Gavin Hamilton Green
1862-1961

1862   Born April 8 in an 'old log house', Lot 11, Concession 12, Colborne Township, Huron County.

1865   Moved to the village of Dungannon, Northeast of Goderich about 20 kilometres.

1868   Enrolled in a private school in Dungannon operated by "a maiden lady of the name of Green (no relation)" at a tuition fee of 10 cents per week.

1870   Transferred to the old Dungannon Public School, a frame building at the site of the current brick structure, County Road 1 just North of the village.

1873   Moved to Port Albert on the Lake Huron shore, 15 kilometres North of Goderich.

1874   Moved to Goderich and attended the old Central School, site of the current Huron County Museum.

1876   Moved to Sheppardton, South of Port Albert, on Highway 21 North of Goderich, where he, as the oldest child, was "boarded out" to do chores for various farmers to save on family rations after a particularly hard winter. He also lived with an uncle in the butchery business in Tiverton, Bruce County, for a short period at this time.

1877   Returned to Sheppardton and went to work in a rake factory. Loaned out to "an old lady" in Slabtown, now Saltford – across the Maitland River by Goderich.

1878-80   Finished final years of schooling in the off-seasons at Sheppardton, graduating with "Third Book honours, if such there be."

1880   Left his familiar stomping grounds of Colborne Township in Huron County to seek his fortune in the American and Canadian mid-West.

1883   Worked for a telephone company in Saginaw, Michigan stringing telegraph wires between Philadelphia and New York. Also spent some time stringing electric light wires in Winnipeg.

1884   Farmed for a short period near Carberry, a small village near Brandon, Manitoba.

1886   Served as a hand on the C.P.R. passenger and cargo vessel the 'Alberta' on Lake Huron and Superior "between Owen Sound and the American Soo," follow ing which he returned briefly to wire stringing in various American States.

1888   Spent a brief period back in Sheppardton working at a sawmill.

1892   Joined the Western Star Lodge of the International Order of Oddfellows at Carberry and remained a member until his death in 1961, at which time, he was Canada's oldest Oddfellow.

Returned from Manitoba and married Agnes Martha Bogie on December 21 at Eagle's Cliff, the residence of Captain Andrew Bogie, Lot 13, Lake Road West, Colborne Township, near Sheppardton.

1893   Set out in search of "a bright future on the matrimonial sea" by returning to Carberry to open a second-hand 'antique' shop which was later destroyed by fire.

1900   Returned in June to his "birthplace on the banks of Lake Huron" citing ill health and his doctor's advice to settle where "the bird's songs seemed to sing `Home Sweet Home', Gavin."

1902   Opened business on Hamilton Street, Goderich "as a dealer in antiques, curios, and used furniture."

1924   Acted as Chair for the founding meeting of the Goderich Historical Society.

1925   Forced to re-locate his business across the street when the two-storey clapboard building burned in September.

1927   Contributed to various historical articles published in The Signal of July 28 to commemorate the Centennial of the Town of Goderich.

1933   Published one of his early newspaper columns "Reminiscence and Lament by the Old Town Clock," reflections of days gone by in Goderich and Huron County. This column and others were collated to form the basis of his manuscript for "The Old Log School."

1939   Published his first book, "The Old Log School and Huron Old Boys in Pioneer Days."

1948   On May 13, he was forced to cancel an auction sale at his shop, already in progress, due to a lack of bidders for his "rare Huron curios" collected over half a century. On May 30, closed the Olde Curiosity Shoppe at 77 Hamilton Street, Goderich.

Published his second book, "The Old Log House and Bygone Days in Our Villages."

1950-53   Collaborated with longtime friend, J. H. Neill, to contribute selected artifacts from his own collection to the basic holdings of the newly established Huron County Pioneer Museum.

1954   Connived, in a good-natured fashion, with J. H. Neill, then Museum Curator, and friend, Harry McCreath, to acquire, repair and 'sell' to the Museum its highly prized Orchestral Regina Organ, presently dedicated to the memory of Gavin's aunt, Charlotte Sophia Green, the first white child born in Colborne Township at Gairbraid, near Goderich on April 8, 1835.

1957   Saddened by the passing of Agnes Bogie Green, "wife and helpmate, who guided me aright through sunshine and sorrow during 64 years of married life."

1961   Died, in his 100th year, at Alexandra Marine and General Hospital, Goderich, where he had been in care for his last three years. Buried in Colborne Cemetery, between Saltford and Benmiller, near Goderich.

1963   Commemorated by Huron County Council with a plaque unveiled in his memory at the County Museum.

1992   Honoured through the re-publication of his 1939 work, "The Old Log School."

THE OLD DUNGANNON SCHOOL was re-built in 1872 following a fire which destroyed the former one room log building. This fine brick structure survived to serve the children of Dungannon village and North for almost a century. Shown with teachers Fred Ross and Miss Mary Durnin are the children in attendance in 1925.

EIGHT GRADES TOGETHER IN PIONEER TIMES was the fore-runner of the less complicated yet controversial combined grade classes in today's schools. These pictures show multi-grade groups during the 1890's outside schools in Sheppardton, Ashfield (1911) and Colborne Township. The classes were organized into 'books' according to which level of reader each child had advanced. Gavin 'graduated' with "third book honours, if such there be". Others continued to complete "fourth" and "fifth" book levels before moving on to the Grammar School in Goderich first established in 1841 and followed by the first High School in 1873 which became the Goderich Collegiate Institute in 1892.

CHARLOTTE SOPHIA GREEN was the first white child born in Colborne Township, at Gairbraid, April 8, 1835. Aunt Charlotte helped to revive the two-year old Gavin from his second battle with life following the "knock-out blow" induced by Uncle John Kerr's whiskey.

EAGLE'S CLIFF FARM, was the site of Gavin's marriage to Aggie on December 21, 1892. Enjoying a pleasant summer afternoon at the Bogie farm, as captured by the famed Goderich photographer, R. R. Sallows, are (L. to R.) Captain Andrew Bogie, Gavin with the family dog, May Bogie-Rome with the pet canary, Aggie, Sarah Bogie-Hawkins and mother Martha. (Note the shadow of the cat (R) which was captured in this time-exposure photograph as it paused momentarily to jump down from the porch.)

"SWEETHEART OF MY YOUTH and guiding star" are the opening words of a poem by Gavin Green which he wrote on the back of this picture in 1958 following the death of Aggie one year before. This formal portrait presents her at age 20.

OLD HOME WEEK IN GODERICH in 1952 saw Aggie, age 80, and Gavin, age 90, costumed (and armed with an 1877 rifle!) to commemorate the devastating hunts of yesteryear which depleted to extinction the great flocks of passenger pigeons, once so dense they would darken the Huron skies as they flew past.

Site Reference List

BENMILLER – This picturesque village is easily found from any direction via the main highways and County roads to the East or North of Goderich. Shown as Colborne Mills on an early map, the site was formally named to commemorate the contributions of Ben Miller, an English settler early to arrive in Colborne Township.

BOGIE'S BEACH AND HALEY'S CORNERS – About 6 kilometres North of Goderich on the Lake Shore Road, now known as Highway 21, one enters the area best known by Gavin Green during his early childhood. Bogie's Beach carries the name of early settlers along this section of the lakeshore. Haley's Corners, probably at Concession 11-12, is near to the very farm where Green was "born in the bush" in 1862. It is named, not for any residence thereof, but rather for one of owners of the more-than-a half dozen inns and taverns which were located along the lake shore road in the heydays of early Township development between Goderich and Shepardton, further to the North. Nearby is the Eagle's Cliff Farm (also known as Cedar Cliff) of Captain Andrew Bogie, illustrated in the text, Lake Road West, Lot 13, newly occupied by Lloyd and JoAnne Lockie. Watch for their mailbox, just South of the Lake Huron Resort Road on Highway 21, near Sheppardton.

CARLOW – originally known as Smith's Hill, is located on County Road 25, Northeast of Goderich. The Township Hall (c. 1869) was originally a waypoint inn of excellent repute. It is suggested that Gavin and his brother David assisted with the brick laying during the hotel's construction.

CRACKIE'S CORNERS – is found at the first sideroad East of Highway 21 beyond Dunlop, North of Goderich. The brown, insulbrick-covered building at the corner shrouds within its run-down walls, the remnants of the "Ploughboy's Inn", the only surviving structure from the once bustling village of Loyal. (Also Cracky's & Crackey's)

DUNLOP – also known as Millburne and Millbrook, is located at the terminus of County Road 25 and Highway 21 just North of Goderich. The old frame structure on the Southeast corner (marked in 1992 as the Exchange Hotel) was moved from the West of the 'Dennis Steep Antiques' barn when the highway was widened in 1960. It originally faced the Crown and Anchor Hotel on the opposite corner. The original town plan, as shown in 1879, included the full first row of farm lots running South from 'Millburn' (sic) corner to Goderich and was also called Bridgendplace.

DUNGANNON – North of Carlow and Nile, on County Road 1, has survived well over a century of change yet retains its rustic village nature if not the busy, thriving character of shops, hotels and traffic of yesteryear. The site of Green's first schooling (on the 1992 property of the Huron & Kinloss Municipal Telephone Co.) – on Main Street across from the former Anthony Black's Hotel or the Old Prince of Orange (currently the offices of the West Wawanosh Mutual Insurance Company) – and later just North of the village in a large frame structure which preceded the now-converted brick school, is worth a visit.

No evidence remains today of a very unpopular tollgate once found South of the village towards Nile, near the former Barr Presbyterian Church at Conc. 3, Lot 12, Ashfield Township. As well a dastardly toll bridge could be found at Glenn's Hill, North of the village towards Lucknow. (Tollgates were finally banned by Huron County Council in 1873!)

After you explore the village, head West on the paved road towards Port Albert and the lake to drive down DISHER'S HILL. Stop at the bottom to view the remains of the 19th century raceway and dam on the Nine Mile River for the original Disher Carding Mill, close to the site of the current Dauphin Feed and Supply depot, downriver from the Saunby Grist Mill, one mile and a quarter East of the former village of Cransford. As you approached the cemetery at the crest of the hill you drove by the site of Green's first Dungannon home (South side) at the western edge of the village. Proceed westerly past the bridge and up the hill. You will arrive at the abandoned site of where young Gavin lived in "a big house" owned by John Runciman, his father's employer at the now-forgotten Runciman Saw Mill. On the North side of the road at this location was the pioneer farm of Richard West.

GAIRBRAID – The 1840s townsite envisioned by the Dunlops – the famous 'Tiger' and brother Robert of Canada Company fame – was located immediately North of the Maitland River at Goderich, and divided into two blocks by the Old Highway 21.

Drive the Tiger Dunlop Road just North of the river bridge and climb the hill to the heritage tombsite for a splendid view of the lake and the river valley. It is said that many buildings were located in this area until the lure of less restrictive United States land development policies led to the abandonment of many farms along present-day Highway 21 North to Dunlop's Hill. Alas, what has changed!

'Gairbraid' was also the name of the rambling, long gone, H-shaped Dunlop log home over-looking the river valley.

GODERICH – The "old town clock" as described in Green's "reminiscence and lament" no longer watches down from its domed tower over the Goderich folk, and, much has changed in the half century since the boardwalks around the Court House Square were lined with visitors to greet the Governor General in 1933. Gavin's extensive inventory of buildings and sites is more meaningful to today's reader if one re-traces the short trip from the former CNR station – now the terminus of the 1992 RailTek "Goderich and Exeter Railway Company" – at the top end of East Street, to the Square and thence to the harbour area. While many of the structures identified in the 1933 account have long since disappeared, one can sense the majesty of the former streetscapes and can easily assume the unabashed pride with which Green described the old port town of his Curiosity Shop days.

Beginning at the RailTek station, over-looking 'Slabtown' on the river flats below, one can peer to the 'left' to find the present day site of Champion Road Machinery, the home of the "Good Roads" graders that are still shipped all over the world.

In the second block of East Street, proceeding to the Town Square, the East End Gym and the National Shuffleboard Company occupy the former "old knitting factory" (Holeproof Hosiery). The McCallum and Palla Funeral Home minds the space of the former world famous Goderich Organ Factory. Further along, the "Oatmeal church" has long since been replaced by a contemporary structure, while the Victoria Street United, the Baptist (on Montreal Street) and the North Street trio of United, St. George's and St. Peter's churches have all survived the ravages of time.

In the final block approaching The Square, the Canada Post facility now holds dominion over the site of the majestic old three-storey Town Hall – complete with its former tunnel to the back alleys where afternoon paper boys once gathered bundles of the Toronto Telegram. The rest of that streetscape, save but for a few South side structures, has been replaced by not-so-lofty 20th

century counterparts to the old high front brick and stone structures first erected in the 1870s.

For the record it is worth recalling that the old courthouse in
the centre of The Square, built in 1856 and destroyed by fire in
1954 was replaced by the then ultra-modern limestone-faced
structure in 1956. Like other 19th century business sections in
small towns, the Goderich Square was not excused from the ravages of fire. As early as 1873 major destruction (13 stores in one
conflagration!) or as recently as the demise of the British Exchange
Hotel (site of the Woolworth Store today) in 1956, there is ample
evidence of one of the primary causes for Main Street architec-
tural renewal.

Sites worthy of note which have not changed since Green's
walkabout with the Governor General include, as we turn onto
the Square at East Street, the facade above the Big V Drugstore, c.
1906, formerly the People's Store, and the buildings comprising
the Bricker Jewellery Store and Baechlar's Kitchen Centre were
certainly present in this same form in the 1930s.

From Hamilton Street to North Street, the facades are currently undergoing dramatic change in an attempt to restore the
heritage flavour that would have been noted a half century ago.

The Anstett Jewellery renovation at the corner of North
Street is the site of the former Sam Detlor Clothing Emporium
and the outlet for the Atlantic and Pacific Tea Company grocery
chain. The upper storeys of the block thence to Colborne Street
are largely intact from their original 1800s construction.

The current Canadian Imperial Banking Centre was next to
Ball's Restaurant; Pridham the Tailor occupied a portion of the
Rivett's store and the accounting firm framed by the Ionic
columns in the red brick structure at the corner of West Street,
once the Bank of Montreal, stands below the only remaining portion of the stately second floor Acheson Opera Hall, c. 1875.
Several of these enterprises are mentioned in the Green text.

Proceeding beyond West to Montreal Street, only the 1882

Craigie building – the Goderich Entertainer – reminds us of the former three-storey streetscape profile. Was it in this small segment, tucked into a second floor studio, that the famed Reuben Sallows did his trade in tintypes, glass negative prints and world famous portraiture?

In the next block, Woolworth's has replaced the old British Exchange Hotel with its second storey, full front balcony; and, from South Street to Kingston, the Bedford Hotel with the (Polley) Livery behind now occupy space formerly graced by the twice-doomed and fire-ravaged Albion Hotel of the mid-1800s.

At the Kingston Street corner, we can still view the second storey front of the Victorian Opera House, but no street level facade from the 1930s reminds us of the liquor store, the tailor shop and the shoe shop present at that earlier time.

Travelling on West Street towards Harbour Hill one must certainly note the grand 1913 Masonic Temple, the former Dominion Post and Customs Office – the current Town Hall at 57 West – and, closer to the hill, the Park House, once the offices of Thomas Mercer Jones, Commissioner of the Canada Company.

At the commercial harbour, the Western Canada Flour Mills are long gone, but the Goderich Elevator and Transit Company still thrives today. Ship Island has long been dredged away but we are still not certain whether Bill Forrest ever got his compensation, as queried in Green's textual challenge to the Governor General.

GREEN'S GODERICH HOME – Gavin made the daily trek from his humble 1 1/2 storey frame home at 86 Anglesea Street (formerly numbered 46) to light the lamps in his non-electrified Olde Curiosity Shoppe at 77 Hamilton Street for almost 50 years. Both buildings, now clad with aluminum siding, are easy to find.

LEEBURN – has long since disappeared, but if you drive North along Highway 21 to the Sunset Golf Course, you will be at the heart of the former community. Two 'kirks' were built on the corners – the

Bible Christian and the Presbyterian (c. 1875, but destroyed by fire in 1878) – on lands donated by the Strachans. Following the fire, by 1880, John Linklater offered land for the new Leeburn Methodist structure on the opposite corner. Two blacksmith's shops, a mill and machine shop, a temperance hall and several residences made up the Leeburn village site, named after the original Strachan homestead near Glasgow in Scotland.

MENESETUNG PARK – was a tiny settlement, near Gairbraid, on the bank of the lake behind the present day Sky Harbour Airport. It is comprised largely of cottages, one being built by an American from Detroit, a Mr. Fry with lumber and furnished with pews from the old Shepardton Church (c. 1882) "where saints and sinners went for over fifty years", when it was sold off in 1932 by auction for scrap. Its main attraction from 1895 until a fire in 1936 was a two storey, frame, 18-room family hotel.

MILLBURNE – see DUNLOP

POINT FARM, Wright's Point or Four Mile Point – just that far North of the County Town at the current site of Point Farms Provincial Park – originally boasted the old Farm Branch Tavern at the highway and a massive three-storey frame hotel (c. 1870) at the beach for up to 200 guests.

PORT ALBERT – located some 15 kilometres North of Goderich at the mouth of the Nine Mile River presents a charm reminiscent of days gone by. Only a few of the 19th century buildings remain. Check the Inn at the Port and the General Store by the bridge at the river's edge. Explore the contemporary fish ladder of recent date and search out the locals for particulars of heritage interest. In Green's time there was an old iron bridge, and opportunity to use a perilous ford to cross the river. Be sure to read the plaques on the two commemorative cairns along Highway 21. One of these relates to the old St. Andrew's Church recounted in Green's text.

In the old days, this 'town' was a busy lake port and home for mills galore: grist, saw, and shingle. A blacksmith shop, a leading factory (serving the schooner fleets?), hotels, post and telegraph office and school houses combined to create a self sufficient community nourishing the needs of nearby farm settlements. The old schools (c. 1841-47) were on London Road at the South end of the village.

SALTFORD – also known as Maitlandville and Slabtown (and the Heights which are originally part of the eastern side of the Gairbraid Blocks), sits below Goderich on the Maitland River flats. In the latter 1800s as many as seven salt wells flourished here, as did a cooper shop, two taverns, two general stores, a pair of tanneries, a brewery and hop yard, a lime kiln, a sawmill and a blacksmith shop. A wooden bridge built in 1856 was the first dependable fixed structure to join the flats to Goderich town. The remains of the 1883 abutments for the first iron bridge can be seen today.

It was here that young Gavin was loaned out to the "old lady" from whom he snitched the "strippings and the cream", and sampled much too much of her home brewed wine!

SHEPPARDTON – also Shepardton, about 7 kilometres North of Goderich was named for the first settler in the area. The white frame building on the highway curve was surely one of the old General Stores moved near the site of the burnt out Royal Hotel. While a few ancient outbuildings may be found in the area the only obvious relic in 1992 is the old tin-sided school just East of the hamlet. There are no signs of the former Morrish or Johnston Mills, the rake factory, the churches, the stores, the post office and the blacksmith shop which made up this townsite.

Glossary of Selected Out-of-Use Words and Phrases

alum
a crystal commonly used as a medical astringent or styptic, also in making pickles

carding
the act of combing wool

coopering
barrel making

Dall
a local substitute for 'Damn'

depokerate
a localism, nonsense word as to discombobulate with force (see text)

derry
a coarse cloth, presumed to be like denim

deviltry (sic)
devilry, wicked mischief

'did the grand'
made a great impression; perhaps 'did the honours'

drawknife
a double-handled knife grasped by two hands and drawn towards the operator to shape lengths of wood which are secured in a 'horse'

duck
coarse cloth or canvas

fain
(see memorial poem) inclined towards, to have a tendency to

forninst
in front of; up against; next to

frow
a wedge-shaped tool for splitting wood

gad
(n.) a long narrow stick used to inflict a sting; (v.) to ramble about idly

gaiter shoes	shoes covered by a cloth fitting over the leg; a spatterdash
histrionic	dramatic, theatric
itch	a skin disease resulting from a tiny species of mite
pannikan	small pan or cup
phaeton	an open, four-wheeled carriage, often drawn by two horses
'play hob'	to be a companion with
precentor	the leader of a choir or the person who leads the psalms in a church
scrip	a note to be exchanged for goods or money
senna	leaves of Cassia plant; used as a tea or laxative
shinny	a game played with ball and clubs, like hockey
sig, sigged	to signal; to set a dog upon something
siller (sic.)	silver
skein	a hank of yarn
skidway	an incline on which to slide logs
shingling hair	to cut in layers with scissors
so 'mote' it be	so 'may' it be
staves & headings	thin pieces of wood forming the sides of a barrel and the pieces shaped to form the top or bottom of a barrel, keg or bucket
tanbark	the bark of willow, oak or other trees used in tanning leather
traces	the harness straps by which a carriage is drawn
treacle	sugary syrup; like molasses
weal	well-being
whiffle tree	a pivoted cross bar at the front of a wagon to which horse harness is attached, also whipple tree
wincey	a durable cloth composed of a cotton warp and a woollen weft; like flannel
Zouave	warrior (origin unknown)

Rock Me To Sleep Mother

Backward, turn backward
O time in your flight,
Make me a child again,
Just for tonight.

Mother come back
From that echoless shore,
Take me in your arms again,
As in days of yore.

Kiss from my forehead,
The furrows of care,
Smooth the silvery strands
Out of my hair.

Over my slumbers
A silent watch keep,
And when my earth-gate closes,
Mother, rock me to sleep.

Rock me to sleep.

*P.S. The Bible says to honour your father and mother. Gavin H. Green
says honour your mother first who brought you into the world, who fed you
from her gentle breast, upon your cheeks sweet kisses pressed, who wiped
away the tears when trouble and sorrow crossed our path and guided us
through childhood days.*
*Father brought home the bacon. That is about all most fathers did to help
to lighten the load of mothers that nursed us through childhood.*

May God bless all mothers.

Gavin H. Green, August, 1957 (Age 96).